Private Firms and Public Water

Realising Social and Environmental Objectives in Developing Countries

Edited by

Nick Johnstone and Libby Wood

International Institute for Environment and Development, London, UK

Edward Elgar
Cheltenham, UK • Northampton, MA, USA

Published by
Edward Elgar Publishing Limited
The Lypiatts
15 Lansdown Road
Cheltenham
Glos GL50 2JA
UK

Edward Elgar Publishing, Inc.
William Pratt House
9 Dewey Court
Northampton
Massachusetts 01060
USA

This book has been printed on demand to keep the title in print.

A catalogue record for this book
is available from the British Library

Library of Congress Cataloguing in Publication Data

Private firms and public water : realising social and environmental objectives in developing countries / edited by Nick Johnstone and Libby Wood.
     p. cm.
   Includes bibliographical references and index
   1. Water utilities—Deregulation—Developing countries. 2. Water supply—Developing countries. 3. Water utilities—Environmental aspects —Developing countries. 4. Sewage disposal—Deregulation—Developing countries. 5. Sewage disposal—Environmental aspects—Developing countries. I. Title: Realising social and environmental objectives in developing countries. II. Johnstone, Nick, 1965– III. Wood, Libby.

HD4465.D44 P74 2001
363.6'1'091724—dc21

00–047637

ISBN 978 1 84064 587 3

Printed and bound in Great Britain by
Marston Book Services Limited, Didcot

Contents

List of Contributors

Nick Johnstone
Pollution Control and Prevention Division
Environment Directorate
OECD
Paris 75016
France
Tel   (33) 1 4524 7922
Email   Nick.Johnstone@oecd.org
(Formerly Research Associate, Environmental Economics Programme, IIED)

Libby Wood
Assistant Project Manager
Mining, Minerals and Sustainable Development, IIED
1A Doughty Street
London, WC1N 2PH, UK
Tel   (44) 20 7269 1630
Email   Libby.Wood@iied.org
(Formerly Research Associate, Environmental Economics Programme, IIED)

Robert Hearne
Professor – Environmental Economics
CATIE
Apdo. 7170-CATIE
Turrialba
San Jose, Costa Rica
Tel   (506) 5568514
Email   Rhearne@catie.ac.cr
(Formerly Research Associate, Environmental Economics Programme, IIED)

Sergio Mazzucchelli
Director, Environment Area
IIED- América Latina
Av. Gral. Paz 1.180
Capital Federal
Buenos Aires, Argentina
Tel   (54) 11 47021495
Email   sergio.mazzu@sei.com.ar

Martín Rodríguez Pardinas
CEER - Instituto de Economía
Universidad Argentina de la Empresa
Lima 717
1073 Buenos Aires, Argentina
Tel   (54) 11 43797588
Email   marp@impsat1.com.ar

Margarita González Tossi
Associate Researcher
Environment Area
IIED- América Latina
Av. Gral. Paz 1.180
Capital Federal
Buenos Aires, Argentina
Tel   (54) 11 47021495
Email   iied-ma@sei.com.ar

Cristina C. David
Philippines Institute for Development Studies
NEDA sa Makati Building, 106 Amorsolo Street
Legapsi Village Makati City,
Manila, Philippines
Tel   (63) 2 8121478 or (63) 2 8939579
Email   cdavid@pidsnet.pids.gov.ph

Lilian Saade Hazin
IHE - Delft
PO Box 3015
2601 DA Delft
The Netherlands
Tel   (31) 15 2151770
Fax   (31) 15 2122921
Email   saade@ihe.nl or lsaade@iname.com

Aké G.M. N'Gbo
Centre de Recherches Microéconomique du Développement
Université de Cocody
Abidjan, Côte D'Ivoire
Tel   (225) 22445915 or (225) 22487882
Fax   (225) 22449858 or (225) 22487826

Acknowledgements

The authors are grateful to the Danish Development Agency (DANIDA) for funding this work. The Swiss Development and Co-operation Agency (SDC) also provided valuable funding towards preparatory work. Grateful thanks are also due to the UK Department for International Development (DFID) for providing financial assistance towards a workshop which brought together the case study authors and others. The editors would also like to thank Frances Reynolds and Jo Clayton for invaluable assistance in the administration of the project and preparation of this document.

1. Introduction

Nick Johnstone and Libby Wood

1.1 BACKGROUND

Traditionally, the provision of water supply and sanitation (WSS) services in developing countries has been the responsibility of the state. Substantial private sector involvement was considered inappropriate given the public good and basic need characteristics of water supply and sanitation services, and the belief that monopolistic tendencies were inherent in the sector due to economies of scale in service provision.

However, in recent years there has been a significant increase in private sector participation (PSP) in the provision of water and sanitation.[1] Between 1990 and 1997, the cumulative new private sector capital expenditure in WSS projects in developing countries was US$25 billion, compared with US$297 million in the period 1984-1990. According to one estimate, by 1997 a total of 97 projects had been implemented in 35 developing countries, ranging from management contracts to leases, concessions, divestitures and build-operate-own-transfer (BOOT) agreements (see Silva et al. 1998).

The increase in PSP has been driven in large part by a desperate need for increased capital investment in WSS in many cities. In many developing countries, the combination of rapidly growing urban populations along with a reduction in assistance for WSS services from international development agencies and decades of poor public management of WSS, mean that public sources of finance are no longer able to bear the costs of system rehabilitation and expansion. Until relatively recently, most formal WSS projects in urban areas were based on state-provided, supply-side solutions. It was generally assumed that since most aspects of the sector (collection, treatment and distribution of water; collection, treatment and disposal of wastewater; and sewerage) were characterised by considerable economies of scale, they should be provided by a single authority at a standardised level. Unfortunately, this approach has tended to result in "over-engineered" high-cost solutions, requiring large government subsidies and has been characterised by high levels of inefficiency and low levels of coverage. Given the low coverage, richer neighbourhoods have been the primary recipients of these subsidised services, while poorer households have tended to have to pay the full cost of whatever alternative strategy is used.

Considerable research has been undertaken on the effects of PSP in terms of the potential for increased efficiency *versus* the potential for anti-competitive monopolistic behaviour, but very few studies have looked at the potential for market failure related to social and environmental concerns. This study is concerned, therefore, not with the relative merits of private *versus* public sector provision of urban WSS in developing countries *per se,* but rather with how best to realise social and environmental objectives when there is increased PSP. In particular, the study looks at the ways in which PSP impacts on the realisation of social and environmental objectives and, in turn, the role that the state should play in achieving these objectives given the existence of PSP. In addition to a review of relevant literature and a two-day workshop, the study draws on the experiences and lessons learnt from five cities: Manila (Philippines), Buenos Aires and Córdoba (Argentina), Abidjan (Côte d'Ivoire) and Mexico City (Mexico).

There is little question that badly regulated PSP poses a threat to all users (including poorer households), granting a private firm monopoly rights to service provision. With costs of provision often higher and levels of demand lower in poorer neighbourhoods, private firms are unlikely to have sufficient incentive to improve access in these areas if they are left to their own devices. However, if regulated effectively and if the service provider is innovative, PSP need not necessarily be a threat to poorer households and neighbourhoods. Indeed, examples such as Buenos Aires, La Paz and others show that provision for the poor can improve with PSP, but this also requires concerted and well-directed effort from all parties. Similarly, environmental benefits may also be realised since increased investment levels may reduce environmental pressures and since the state is no longer faced with the contradictory incentives of being both provider and regulator.

The research findings illustrate that although the state is no longer the direct provider of WSS services, its role remains key. Due to the very "public" nature of WSS provision, the state should ensure that the externalities associated with adverse effects on public health and the environment are internalised and that aggregate water use is sustainable and efficiently allocated. In addition, it should ensure that household consumption meets minimum acceptable levels in terms of quantity and quality. To ensure that social and environmental goals are realised, they must be prioritised throughout the process, from the initial stages of contract design to the selection of technologies and management solutions.

The remainder of this introductory chapter summarises the key issues to be discussed in the report, including key social and environmental concerns, arguments for PSP in WSS and a description of the forms it can take, and the institutional characteristics of the sector. Chapter 2 discusses the social and environmental implications of the shift from public to private service providers. Particular attention is given to the changing role of the public authorities. Chapter 3 discusses WSS provision in low income neighbourhoods and the scope for service differentiation and decentralised management. Chapters 4 to 7 describe the experiences with PSP in WSS in the four case study countries. Chapter 8 draws together the main conclusions, focusing on the policy implications of PSP with respect to social and environmental objectives.

1.2   SOCIAL AND ENVIRONMENTAL CONCERNS

There is an urgent need to ensure that poorer households in urban areas gain access to affordable WSS. Although the proportion of urban dwellers with piped water and sanitation provision in developing countries has increased over the last 20 years, the rapid increase in urban populations has resulted in the total number of people unserved actually increasing. According to UN figures, in 1994, almost 300 million urban dwellers in developing countries were still not served by water supplies and almost 600 million were without sanitation. However, these UN statistics almost certainly exaggerate the number of urban dwellers that are "adequately" served.[2]

There is some regional variation in the proportion of urban populations served by water supply and sanitation. Africa reportedly has the highest proportion of unserved urban inhabitants, with 36 per cent for water and 45 per cent for sanitation. Conversely, the situation in Latin America and the Caribbean is much less acute, with only 12 per cent and 27 per cent of the population without access to water supply and sanitation respectively (see Table 1.1).

Not surprisingly, it is almost always poorer households who do not have access to adequate water or sanitation provision. For instance, in Peru approximately 70 per cent of higher-income households have access in contrast to only 10 per cent of lower-income households (see UNICEF/WHO 1996, and the Manila and Buenos Aires case studies).

Table 1.1: Coverage of WSS for Urban Areas by Region (1994)

Region	Urban Population Served by Adequate Water Supply		Urban Population with Sanitation Coverage	
	Millions of Inhabitants	Percentage	Millions of Inhabitants	Percentage
Africa	153	64	131	55
Latin America & Caribbean	306	88	254	73
Asia and the Pacific	805	84	584	61
Western Asia	51	98	36	69
Total	1,315	82	1,005	63

Source: UNICEF/WHO (1996)

Excluded households and neighbourhoods have adopted a variety of alternative informal strategies in an attempt to ensure levels of provision necessary for a reasonable level of health. In some cases these strategies, while being unregulated and unlicensed, are similar to those adopted by some small-scale formal systems. However, they are often quite different in terms of their management (see Altaf 1994 and Solo 1998 for discussions). For water, these range from direct reliance on surface- or groundwaters through tubewells, to water vendors and illegal connections. For sanitation, households may rely on simple pit latrines, toilets connected to septic tanks, drainage canals, and in some cases even simplified sewerage systems (see Altaf 1994 and Solo 1998). These informal systems may be managed at the level of the household, but "collective" measures are also common. For instance, it is not uncommon for groups of households to form associations to share the cost of sinking wells and managing the supply, or to manage multiple household septic tanks. Commercial ventures are also common, with the majority of residents in many cities relying more upon unlicensed vendors than formal systems, whether public or private.

While some informal systems are appropriate and efficient responses to the absence of formal service provision, in many cases the strategies adopted by households, neighbourhoods and small private operators are short-term solutions to the problem of inadequate access. These strategies may not be efficient and are likely to have significant negative social and environmental consequences.

To a large extent, the realisation of social and environmental objectives are complementary. For instance, increased access to adequate WSS facilities for the urban poor is likely to reduce excessive groundwater abstraction and seepage of wastewater. However, they can also be contradictory. Investment funds directed towards higher levels of wastewater treatment may come at the expense of service expansion into poorer neighbourhoods.

Social Concerns

This lack of access to adequate and affordable WSS has numerous interdependent consequences for poorer households:

- **Increased monetary costs for those who lack access**. As noted above, when households lack access to formal centralised systems (whether private or public), they adopt a variety of alternative strategies to try to meet their needs. In the majority of cases, these alternatives are more expensive than formal systems. Thus, access to WSS is an important determinant of a household's level of welfare.

- **Increased time and physical effort needed in collecting water**. Fetching and carrying sufficient water for a household's domestic needs is a time-consuming and arduous task, even from a standpipe 20 to 30 metres away from the home. The time spent obtaining water may limit income-earning opportunities.

- **Reduced water consumption levels**. The greater the cost - in time, effort and money - of water and sanitation, the less likely households are to adequately cater for all domestic needs. Indeed, households lacking water supplies piped to their home or yard tend to have much lower consumption levels. For instance, average monthly *per capita* consumption rates for water in Manila are $5.8m^3$ for households with access to piped water, but only $1.3m^3$ for households which rely upon public faucets, and between $1.3m^3$ and $2m^3$ for households which rely upon different types of vendor (see David and Inocencio 1998). As a consequence, the richest group of households consumes 15 times more water than the poorest.

- **Increased health burdens**. At any one time, almost half the urban population in developing countries is suffering from one or more of the main diseases associated with inadequate water and sanitation provision (see WHO 1996). These include various water-borne diseases (e.g. mostly diarrhoeal diseases and intestinal worms), water-washed diseases (e.g. various skin and eye infections such as scabies and trachoma), and water-related insect vectors (e.g. malaria and dengue). These diseases arise from consumption of water of inadequate quality and are often exacerbated by inadequate sanitation. Consumption of insufficient water volumes also results in significant adverse health effects.

- **Economic costs in terms of lost productivity**. The adverse health effects of inadequate WSS facilities result in household members taking time off work due to illness or to nurse sick family members.

Environmental Concerns

The consequences of inadequate WSS provision are not confined to those directly affected. The adverse environmental consequences resulting from excessive consumption and inefficient allocation of water and discharges of inadequately treated wastewater often extend outside of the urban area.

In particular, in many cities, excessive consumption by urban users (residential, commercial and industrial) adversely affects competing uses outside the city's boundaries reliant upon water from the same catchment area.[3] Rural uses such as water for irrigation and rural household consumption are affected. The overuse of water within urban areas also affects the water available for ecological services such as the maintenance of wetlands and fish populations.

Similarly, in many cases, excessive groundwater abstraction adversely affects other urban users (through reduced availability and increased pumping costs), and also has associated environmental effects such as land subsidence and saltwater intrusion. There are numerous examples of cities in developing countries where the demand for water is outstripping supply and leading to environmental degradation. For example, in Bangkok, which overdraws water from its aquifer by approximately 0.7 million m^3 per day, the compacting of underlying soils has led to land subsidence ranging from five to more than ten centimetres per year throughout the region. It is estimated that Bangkok would have to reduce its groundwater extraction rate by at last one half to alleviate this subsidence (see WRI 1997). Similarly, depletion of Mexico City's aquifer (which supplies 72 per cent of the city's population) is causing the groundwater level to sink by about one metre per year. Because of this over-extraction, Mexico City is suffering from severe land subsidence (see case study).

The implications for the environmental quality of an area, of the existence of large numbers of households without access to sewerage collection systems can also be significant. Neighbourhoods with inadequate wastewater collection can become severely degraded, particularly if there is poor surface drainage. In addition, with increased population densities many "on-site" facilities may not be adequate. For instance, a household might choose to use a simple pit latrine which is perfectly sanitary in terms of immediate environmental consequences. However, depending upon soil conditions it may result in externalities by contaminating the groundwater supply of the community.

Urban water users also impact upon other water users when they discharge effluent into lakes, rivers, estuaries, and the sea. Indeed, pollution of the urban environment is seen as one of the major obstacles to sustained economic growth. As agricultural and industrial demand for water continues to grow, the progressive degradation of water resources and the inability to treat and recycle water mean that demand is outpacing supply at an increasing rate. As a result of untreated wastewater and the runoff of uncollected wastewater in urban areas, rivers downstream from many cities have high biological oxygen demand (BOD) levels and bacterial counts. The consequences do, of course, depend upon the characteristics of the receiving waters. For instance, cities which discharge into lakes with relatively low levels of assimilative capacity require high levels of wastewater treatment. For coastal cities, maritime discharges are often much less damaging, but can still lead to polluted beaches and reduced fish populations. For example, in Manila, where two rivers carry vast quantities of the city's sewage into Manila Bay, fishery yields declined by 39 per cent from 1975 to 1988 (WRI 1997).

The negative health effects of untreated water may also be experienced outside of the urban area by communities relying on the same rivers as a source of drinking and irrigation water. For example, the Tiete River downstream from São Paulo, Brazil, is heavily contaminated by the city's wastes, yet it is used as drinking water by several rural communities in the interior of São Paulo state and as a source of irrigation for nearby vegetable farms (WRI 1997).

1.3   THE CASE FOR PRIVATE SECTOR PARTICIPATION

The evident failure of past approaches to WSS supply has spurred a number of interdependent shifts in the mainstream/orthodox philosophy of provision in urban areas in recent years. These have resulted in a trend towards increased PSP, more decentralised management, an emphasis on demand-based provision (and differentiated levels of service) and a greater degree of cost recovery.

The case for PSP in the WSS sector stems in part from a belief that the private sector is better placed to undertake the kinds of investment necessary to expand and rehabilitate the infrastructure. The wish to relieve burdens on public finances is not, however, the only factor. Indeed, in many cases it is felt that public authorities have been unable to manage urban WSS efficiently and to undertake the investment necessary to provide adequate levels of service provision. Expansion of the service network has not kept up with demand; those households which are connected face considerable unreliability in service provision due to inadequate maintenance; cost recovery has been negligible; and large financial transfers from government treasuries have been required. A number of closely related reasons have been frequently cited in the literature as being responsible for these problems. [4]

- **Gamekeeper-poacher problems.** Public water and sewage utilities will tend to be inefficiently managed since governments have multiple objectives but limited financial resources. With the government as both owner and provider, the manager of the utility is subject to a number of conflicting influences, which it may not be able to balance if clear priorities are not established.

- **Flexibility and autonomy.** At the level of operations, public utilities are often constrained by bureaucratic requirements which do not affect private firms to the same extent. For instance, there is often considerable inflexibility in the management of human resources within public utilities.

- **Absence of competitive discipline.** Since public utilities are not usually subject to the disciplines of the market they have fewer incentives to minimise costs (and maximise tariff collection rates), and provide services in a manner which customers demand.

- **Access to capital.** With government budgets strained, most public utilities have insufficient financial capital to undertake the necessary investments to maintain (let alone expand or improve) services. It is argued that private companies are better placed to access capital, both domestically and internationally. They may also be better placed to access technical skills (human capital).

However, it must be recognised that there are numerous examples of efficiently managed public water and sanitation utilities in developing countries (see Ingram and Kessides 1994 and Nickson 1996). There are documented cases in countries as diverse as Ecuador, Chile, Zimbabwe and Botswana. Moreover, it is important to bear in mind that the characteristics of the public sector differ between countries. Thus, in most developing countries, it is not necessarily the public sector *per se*, but factors such as faulty incentive structures, the politicisation of personal appointments and management, and other bureaucratic weaknesses that contribute to poor performance.

Concern about "government failure" is not, in itself, sufficient to justify private sector involvement in the sector. A private monopolistic service provider may well exacerbate the situation, taking advantage of its privileged position in the market at the expense of service users. Thus, perhaps an equally important motivation for the increase in PSP is the fact that there is now a widespread belief that PSP need not result in monopoly profits. Moreover, the consensus that WSS services are natural monopolies has been brought into question. While the entire chain of provision (raw water supply, water treatment, water distribution, wastewater collection and wastewater treatment) may exhibit economies of scale, individual aspects of the WSS system may not be characterised by decreasing costs, thus allowing for vertical "unbundling" of the sector with multiple providers at some stages.[5] For instance, metering, operations and maintenance, billing and a number of other aspects of service provision can be hived off from the core of the WSS sector. Similarly, firms can be given contracts to undertake significant investments in infrastructure such as wastewater treatment plants.

In addition, it has been recognised that even if technological and economic characteristics are such that monopoly provision is "efficient", more comprehensive forms of PSP can still be introduced. A single provider will not necessarily behave monopolistically as long as there is competition *for* the market, if not directly *in* the market. Thus, firms can be granted the right to provide all WSS services within a given area over a pre-defined period of time. However, as we will see in Chapter 2 this requires effective regulation.

Despite this widespread belief in the potential to allow for the efficient use of the private sector in some areas of service provision, firm empirical evidence of the relative merits of private and public management of the sector is relatively limited.[6] (Noll et al. (forthcoming) find positive changes in consumer surplus following the introduction of PSP in Buenos Aires, Lima, Santiago and Conakry). This is due in part to comparison being made difficult by the wide variety of forms of PSP and to the fact that PSP is new in most developing countries. Indirect evidence can be obtained through comparison of tariff rates before and after PSP. For instance, in the case of Buenos Aires, tariffs charged by the private consortium (Aguas Argentinas) were 73 per cent of those previously charged by the state utility.[7] In Manila, tariffs in the east of the city and the west also fell (25 per cent and 57 per cent respectively) relative to pre-concession rates. Gabon experienced a drop in tariffs (82.5 per cent) with PSP as well (see Haarmeyer and Mody 1998).

However, to the extent that these figures do not reflect the effects of accompanying reforms in the sector as well as changes in service quality, such evidence is not conclusive. Moreover, in some cases (e.g. Manila) tariffs were raised just prior to privatisation. Thus, more comprehensive evidence can be obtained through examining various efficiency indicators. One of the few studies to look at the issue more systematically in a developing country is Chisari et al. 1997. They found that in the case of Buenos Aires unaccounted-for water fell by 6.12 per cent and purchases of intermediate inputs fell by 4.86 per cent between 1993 and 1995. Labour productivity fell initially, but then rose by 75 per cent.

1.4   TYPES OF PSP

Until relatively recently, the only countries with a significant degree of private participation in the WSS sector were France and the United States. Indeed of all public utilities, WSS was the sector in which the formal private sector was least active and, in developing countries, almost non-existent. However, this is changing rapidly as many developing countries involve the private sector (usually European or North American firms in partnership with a local firm) in the provision of WSS, particularly in urban areas. There are many different options for the participation of the private sector. These can be classified as follows (see Table 1.2).[8]

- **Service contracts.** These are the simplest forms of PSP, involving short-term contracts to provide limited services, such as reading meters, repairing leaks, and mailing statements for payment. These contracts entail carrying out specific duties and do not require any overall private sector responsibility for system operation. (The case of Mexico City (Phase 1 of PSP), discussed in Chapter 5, provides an example of a service contract.)

- **Management contracts.** These require somewhat greater private sector responsibility, with the company assuming day-to-day responsibility for system operation and maintenance. However, management contracts do not require any private investment, the private company does not assume commercial risk, and does not have any direct legal relationship with the consumer. The national or local government must maintain financial responsibility for the system and the capacity to plan and finance system expansion. The contractor will not get paid unless fees are collected from the consumers, and has an incentive to improve system management under the stipulations of the contract. Management contracts can be the first step in initiating a process of more comprehensive PSP. (A management contract is currently being implemented in Gdansk, Poland.)

- **Leases**. These allow a private operator to rent facilities from the public authority for a stipulated period of time. The public authority retains ownership and responsibility for system finance and expansion, but the private contractor accepts some commercial risk in the day-to-day operation of the system. The private contractor is not responsible for any capital costs, and rental fees are often based upon the costs of debt service for capital costs. The contractor has a direct incentive to maximise fee collection since returns are revenue generated less operating costs and rental fees. (Prior to 1987 Abidjan employed this type of PSP. Examples of lease contracts can also be found in Guinea and Senegal.)

- **BOOT contracts**. These are mechanisms to allow a private contractor to *B*uild, *O*wn, *O*perate, and *T*ransfer a specific capital investment such as a wastewater or potable water treatment plant. Usually the investment is quite large and the contract period is long enough to allow for the recuperation of capital expenditure. Generally, the public authority must guarantee a certain demand such as a volume to be treated. The contractor accepts a risk if this demand is not met. There are numerous variations on this option, such as BOTs, "reverse BOOTs" and others (Phase III of the Mexico City contract has been called a ROT (*R*ehabilitate *O*perate and *T*ransfer)). This form of PSP is particularly common in Asia, with major recent contracts in Malaysia and China.

- **Concessions**. These are long-term contracts which require the private company to invest in the system. The concessionaire has overall responsibility for the system, including operations, maintenance, investment and expansion. The concessionaire receives payment directly from the consumer and accepts the risk that costs do not exceed revenues. The contract period is usually long enough to allow the contractor to recover investment costs. Penalties may be imposed upon the contractor if specific targets or standards are not met. (The case study cities of Abidjan, Manila and Buenos Aires have all introduced this form of PSP. Other well-known examples include Macao, and Limeira in Brazil.)

- **Shared Ownership**. This allows for shared government and private responsibility for service provision, through a separate corporate entity. Generally, a corporate agreement will stipulate private and public responsibilities, including representation on the board of directors and division of profits. Private finance may be facilitated by the establishment of a separate credit rating with support from the public authority. (Abidjan has used joint private-public ownership in the past.)

- **Full Divestiture.** Allows for full private ownership and responsibility under a regulatory regime. Public assets can be sold to private parties and revenue can be generated. (The only major example of this option in a developing country is the Thailand East Water Company. See Rees 1998.)

To one extent or another, all of these options have been introduced in developing countries. However, concessions are the most popular both in terms of number and size of investment (see Table 1.2). Regionally, private sector investment has been concentrated in Latin America and East Asia. Sub-Saharan Africa, on the other hand, accounts for less than 1 per cent of total investment.

Table 1.2: Private WSS Projects in Developing Countries, 1990-97

Type	Projects	Total Investment in Projects with PSP (1997 US$ millions)
Concession	48	19,909
BOT/BOOT	30	4,037
Operations and management	13	n.a.
Divestiture	6	997
Total	97	24,943

Source: World Bank (1998)

Clearly, the sharing of responsibilities between the state and private sector will depend largely on the form of PSP adopted. Table 1.3 summarises the respective allocation of responsibilities implied by the different options, and their usual duration.

Table 1.3: Principal Options for PSP and Responsibility

Option	Asset Ownership	Operations & Maintenance	Capital Investment	Commercial Risk	Usual Duration
Service Contract	Public	Public & Private	Public	Public	1-2 years
Mngmt Contract	Public	Private	Public	Public	3-5 years
Lease	Public	Private	Public	Shared	8-15 years
BOOT	Private	Private	Private	Private	20-30 years
Concession	Public	Private	Private	Private	25-30 years
Shared Ownership	Joint Corporate	Joint Corporate	Joint Corporate	Joint Corporate	Indefinite
Divestiture	Private	Private	Private	Private	Indefinite

Source: adapted from Brook-Cowen (1997)

1.5   THE IMPLICATIONS

The involvement of the private sector has important implications for all those involved with the provision of WSS in urban areas, including users, non-governmental organisations (NGOs), community-based organisations (CBOs), and donor agencies, as well as public authorities and private companies. Most importantly, PSP changes not only the roles of each stakeholder, but also the way in which they relate to and interact with one another.

Implications for Public Authorities

The large increase in PSP in the delivery of urban WSS in developing countries may require more, rather than less, effective public sector intervention in the sector. Since many aspects of the sector are not likely to be truly competitive due to the technological characteristics of service provision, public authorities will have to regulate the sector effectively to ensure that services are not over-priced or under-provided. More pertinent to this study, public authorities will also have to ensure that social and environmental objectives are met. Fulfilling these functions is a complex task, not least because realisation of some of the objectives can undermine others.

In many senses the regulatory functions of public authorities with respect to social and environmental concerns will not have changed appreciably with PSP. Implementing agencies with direct responsibility for the realisation of social and environmental objectives may well serve roles which are indistinguishable from those they fulfilled previously when public authorities had direct responsibility for service provision. However, their incentives will be different. Once the private sector bears some degree of responsibility for service provision, the expectations and behaviour of public authorities may well change.

More importantly, the role of the "economic" or "sectoral" regulator will have changed with increased PSP. Indeed, in most cases the "economic" regulator will only come into existence because of PSP, mainly out of concern for the potential of private sector providers to exercise monopoly power. Thus, through a variety of mechanisms (price regulation,[9] service quality standards, coverage targets, etc.), the regulator seeks to ensure that the service provider does not over-price or under-provide services. However, the behaviour of the economic regulator may have very significant environmental and social consequences. Indeed, in some cases the role it plays may be even more important than that of the agencies with direct responsibility for environmental and social matters. However, its impact, while important, is often indirect and unintentional. Ensuring that there is co-ordination and coherence amongst the various public agencies which affect the sector is one of the main challenges facing the sector with increased PSP.

Implications for Private Sector Service Providers

Increased involvement in the provision of urban WSS in developing countries has given European and North American firms access to important new markets with which they are relatively unfamiliar. Involving local firms has certainly helped many firms understand local conditions better. However, uncertainty and lack of information remain endemic.

Information problems are perhaps most acute in relation to social and environmental concerns. In particular, demand amongst poorer households is only imperfectly understood since in many cases they do not have access to formalised forms of service provision and even where they do, the supply-side approach of the past means that existing services do not necessarily reflect their real preferences. Lack of regularised formal development in poorer areas means that the costs of provision are also inadequately understood. Uncertainty is compounded by the fact that standards and regulations related to environmental concerns are changing, probably, faster than in any other sector.

If firms - local and international - are to play an increasing role in WSS provision, it is in their interest to ensure that environmental and social concerns are addressed. There are two reasons for this. First, the existence of a large number of unserved households in the city will adversely affect the service provider's capacity to meet its customers' needs and any contractual obligations that it may have with public authorities. Since environmental and health externalities are endemic in the sector, providers will not be able to "insulate" their operations from conditions in the wider community. Wastewater runoff and unregulated water abstraction by households in unserved areas will affect the firm's own production costs and its capacity to meet contractual obligations, significantly increasing rates of unaccounted-for water and production costs.

Second, successfully addressing these concerns now will benefit the private sector in the longer term. We are still in the relatively early stages of PSP in the WSS sector in developing countries. If the private sector is to continue to be given more opportunities to provide WSS services in developing countries in the future, it will clearly have to prove itself not only commercially, but also in social and environmental terms.

Implications for NGOs and CBOs

NGOs and CBOs have long been active in urban areas in developing countries, seeking to improve urban environmental quality and ensure that poorer households have access to affordable WSS. In some cases this has involved lobbying for greater investment in public services or, more usually, helping communities to make claims on their own behalf. In many of the more successful examples, they have also been active in helping communities undertake investments themselves. (For examples see Luther 1993, Kinley 1992, Whittington et al. 1992, Lyonnaise des Eaux 1998, World Bank 1994.)

To some extent, these functions will not have changed with PSP. However, NGOs and CBOs may find that they have to change their "message" and their "target audience". Lobbying public authorities for increased public investment is likely to be even less effective than has been the case previously. However, paradoxically, the role played by NGOs and CBOs may become more important with increased PSP for a number of reasons.

- First, irrespective of the form it takes, PSP involves a separation of responsibilities. This can be confusing to users, making it more difficult for them to express their needs and preferences. Collective organisations such as NGOs and CBOs have a role to play in ensuring that investments undertaken by private sector service providers reflect the preferences of users.

- Second, it is becoming increasingly clear that, notwithstanding economies of scale, in many cases different levels of services for different households or neighbourhoods may be more efficient than a standardised service. It is only by working closely with organisations familiar with the needs of the users and with the nature of existing services, that the demand for, and cost implications of, differentiated services can be known with any degree of certainty.

- Finally, without soliciting the inputs of local communities and ensuring their active support, private sector service providers are less likely to be successful in ensuring that targets of service expansion and tariff collection are met. Working with NGOs and CBOs can be a cost-effective way of doing so.

Thus, the involvement of NGOs and CBOs may increase with the shift towards private sector service provision. In some cases, NGOs and CBOs may themselves become active in the provision of certain aspects/stages of WSS service provision. As long as technical criteria are met, PSP has the potential to open up the sector to all kinds of service providers, not just large firms. There may well be aspects of service provision in poorer neighbourhoods to which many of the large firms are not best suited and in which CBOs and NGOs are able to play a cost-effective role.

1.6   THE CASE STUDIES

The study draws on the experience of private sector involvement in four case study countries. The cases were selected to provide a range of experience in terms of the nature of PSP adopted, as well different institutional contexts and different social and environmental concerns. The nature and objectives of PSP in each case are summarised below.

Buenos Aires and Córdoba

After decades of inadequate and inefficient public sector provision, private sector firms were invited to bid for water and sanitation concessions in a number of Argentinian cities in the early 1990s, including Buenos Aires and Córdoba. In the case of Buenos Aires, a 30-year concession for the provision of WSS was won by a consortium of nine companies, Aguas Argentinas, led by Lyonnaise des Eaux. Local firms play a minority but important role in both consortia. Regulatory bodies have been set up, bringing together representatives from different levels of government.

Amongst other objectives (i.e., reduced public deficits, increased investment levels, increased service efficiency) it was hoped that the involvement of private sector firms would result in the attainment of a number of environmental (improved wastewater treatment, reduced unaccounted-for water, reduced groundwater contamination and depletion, etc.) and social (improved access to affordable water and sanitation for lower-income households and neighbourhoods) objectives which the public sector provider had been unable to achieve satisfactorily.

Manila

In 1997, the Metro Manila Waterworks and Sewerage System (MWSS) was put out to tender. The decision to introduce PSP was motivated by a commitment to improve the efficiency of MWSS operations, raise financial resources for investments, and end government subsidies. Prior to privatisation, the MWSS service was characterised by low water pressure, high rates of non-revenue water, intermittent supply and low coverage rates. MWSS services accounted for approximately 60 per cent of total water usage in Manila. The remainder has been obtained from privately-owned wells, causing problems of groundwater depletion.

The form of PSP chosen was a 25-year concession agreement. In order to promote competition and generate yardstick information, the service area was divided into East and West Zones and concessions granted to two different companies. A residual MWSS was retained to carry out limited management and facilitation roles. In addition, a separate regulatory office was established. The two concessions were granted to those companies who promised the lowest tariff levels. Since different rates were submitted, a higher bid price had to be accepted in the West Zone and consequently the price of water differs substantially between the two zones.

In terms of social goals, the concessionaires were required to expand coverage of water supply, sewerage, and sanitation services and to provide a 24-hour supply to all connections by the year 2000. By 2006, a 96 per cent coverage is expected. In low-income areas, the concessionaires are obliged to establish one public standpipe per 50 households. Sewer connection coverage is scheduled to increase from 7 per cent to 62 per cent by 2021. In the meantime, sanitation services (desludging of septic tanks every five to seven years) will be used. The costs of expanded sewerage will be passed on to the consumers.

Mexico City

In 1992, the Federal District Water Commission was created with the task of engaging PSP in the Federal District of Mexico City through multiple, multi-stage service contracts for water supply. PSP forms part of the broader Water Management Programme (1995-2000) aimed at overcoming growing problems in the water supply and distribution system prior to PSP. A system of fixed and highly subsidised tariffs, the absence of a payment culture, and high levels of leakage had resulted in financial problems, and these problems were in turn leading to insufficient investment in the infrastructure and to a deterioration in water services. Moreover, excessive demand was rapidly drawing down the aquifer under the city causing the city centre to sink by several metres.

Given the lack of information regarding the customer base, water consumption levels and network conditions, a phased approach to PSP was considered most appropriate. The Federal District was divided into four contractual zones and, following an international bidding process, contracts were awarded to four separate consortia. The tasks of the consortia were to be accomplished in three phases, with the responsibilities of the consortia increasing over time. The Water Management Programme focuses on two main strategies aimed at reducing lapses and deficiencies in the provision of water services: to implement universal water metering for the assessment of customer bills, and to significantly improve the water distribution infrastructure.

Abidjan

Since Independence, water distribution in Côte d'Ivoire has been carried out by a private company, SODECI. In 1974, SODECI was granted a 15-year lease contract for the entire country, giving it responsibility for distributing water and collecting revenue on behalf of the State. The State remained responsible for investment decisions but soon ran into financial difficulties. In 1987, in an attempt to overcome these difficulties, SODECI was awarded a 20-year concession for the country's water supply networks. The new convention improved the co-ordination between investment and operating needs, with SODECI administering the investment funds itself. SODECI does not have investment obligations but can make small investments. The Government also pays royalties to SODECI for maintaining sanitation facilities. A concession contract is currently being prepared in which users will pay SODECI directly for sanitation services.

Water, sanitation and the environment fall under the authority of a number of different ministries which are involved in devising water policy, controlling the concession, and financing works. Regulation of the water sector is divided between the Water Division and the Ministry of Economy and Finance. A new Environmental Code was implemented in 1996 and National Water Law is currently being drafted.

The water tariff is uniform throughout Côte d'Ivoire. A proportion of water revenue goes towards the National Water Fund (intended to reimburse loans to the benefit of the water and sanitation sector) and the Water Development Fund (administered by SODECI and used to finance social connections, renewal works, extension and new investments). SODECI's revenues are calculated on the basis of operating costs with a contractual margin of 5 per cent for the operator.

Approximately 90 per cent of SODECI's service subscribers could be categorised as economically disadvantaged and consume very low volumes of water. The Government, in conjunction with SODECI, has implemented a socially equitable rate structure. In certain circumstances social connections are free. The number of subsidised connections has increased from 14,681 in 1987 to 30,334 in 1997. However, a proportion of the population remains unconnected to the water network since they cannot afford the connection costs or are unable to save with the regularity required for the three-monthly billing system. In peri-urban areas, SODECI works with retailers who supply water to low income unconnected consumers. Retailers pay a deposit to SODECI and subscribe to an industrial use connection (this effectively represents the privatisation of resale). In general, the retail price is higher than the official price.

NOTES

1. Private sector participation (PSP) in this context refers to large private companies such as Thames Water and Lyonnaise des Eaux. The existence of a large number of smaller-scale private entrepreneurs in WSS, such as water sellers and tank truck suppliers is discussed later.

2. There are two reasons for this. The first is the lack of a universal definition of what is "adequate", and the consequent latitude given to governments in judging who is adequately served. Walking distance from household to water source was the principal criterion for water supply (adequate for the majority defined as less than 50m), and for sanitation, excreta disposal facilities which break the faecal-human transmission route is generally regarded as "adequate". The second is the tendency for governments to greatly exaggerate the proportion of people in their country with access. UN agencies such as the World Health Organization, being inter-governmental organisations, are obliged to publish the statistics supplied by their member governments.

3. As the case of Mexico City illustrates, in many cases urban areas draw water from hundreds of kilometres away, at great cost.

4. See Nickson 1996, Ingram and Kessides 1994, Idelovitch and Ringskog 1995, and Mody 1996 for discussions.
5. See Noll et al. (forthcoming) for a discussion of economies of scale at different stages of service supply (raw water supply, treatment, and delivery).
6. The role of ownership (rather than management *per se*) is better documented, but most of the evidence relates to the United States since it is one of the few countries with a mixed private-public sector. Raffiee (1993) found that the behaviour of public water utilities was much further from cost-minimisation than those owned privately. In an earlier study Crain and Zardkoohi (1978) found similar results, and concluded that most of the observed difference in efficiency was due to differences in labour productivity. Conversely, Feigenbaum and Teeples (1983) did not find that ownership played as significant a role in determining the relative efficiency of water and sanitation utilities as had been thought previously. This result may be explained by the fact that unlike previous studies, they included a number of supply-side variables, which tend to vary systematically with ownership (i.e., source of raw water supply, population density of area served, etc.). They also distinguish between the characteristics of the service provided (i.e., treatment levels, reliability of service, etc.). Teeples and Glyer (1987) refined this analysis, using more general econometric specifications. They found that for the least restricted regression, ownership was statistically insignificant. Two other studies (Lambert and Dichev 1993 and Bhattacharyya et al. 1994) have even found that public utilities are on average more efficient, albeit with much greater variability in performance.
7. They remain 17 per cent below the pre-concession rate, although the tariff structure has just been reformed once again (see Jaspersen 1997).
8. For different taxonomies of PSP and private-public partnerships see Idelovitch and Ringskog 1995, Gentry and Fernandez 1997, and Brook-Cowen 1997.
9. This can either take the form of cost-plus regulation (which guarantees a level of profits for the provider) or price caps (which guarantees a price for users). While the latter is usually advocated since it provides more incentives for efficiency improvements, cost-plus price regulation is more common. However, in practice the two systems of price regulation are quite similar.

2. Regulation of Social and Environmental Concerns with Increased Private Sector Participation in the Provision of Water and Sanitation

Nick Johnstone, Robert Hearne and Libby Wood

2.1 INTRODUCTION

As discussed in Chapter 1, PSP in urban WSS may place more, rather than less, demand on effective public sector intervention in the sector. In some respects this is ironic given that one of the principal arguments for increased PSP is a perception of public mismanagement of the sector. This chapter is concerned with how best to meet social and environmental objectives in light of potential market failures where there is significant PSP in the sector. The discussion focuses on those aspects of the public authority's responsibilities which have changed with PSP, and not on environmental and social policies in the sector more generally. In particular, the chapter looks at the need for the public sector to:

- provide mechanisms whereby aggregate water use is sustainable and allocated efficiently between alternative uses;

- internalise the externalities associated with adverse effects on public health and environmental quality;

- serve as a guarantor of a level of service provision, which is consistent with a basic standard of living.

Following this brief introduction, Section 2.2 provides a discussion of the potential for environmental and social market failures in the sector. Section 2.3 reviews the role of the "economic" regulator in the realisation of environmental and social objectives, and Section 2.4 looks at factors constraining their realisation. Section 2.5 reviews bidding procedures and contract design. Finally, Section 2.6 lists some of the main conclusions.

2.2   SOCIAL AND ENVIRONMENTAL EXTERNALITIES IN THE SECTOR

In addition to the potential for market failure through monopolistic service provision, there are three other types of market failure in the sector which relate directly to environmental and social concerns: over-exploitation and misallocation of raw water supply; health and environmental externalities from wastewater; and the under-provision of basic needs. In Chapter 1, the social and environmental concerns related to water and sanitation were summarised. This section demonstrates how these concerns arise through various market failures.

Conservation and Allocation of Water Resources

Goods that are characterised by low excludability and low subtractability - and are consequently not rationed properly by prices - are public goods. In general where water is scarce and valuable, access to water is characterised by high excludability and high subtractability and thus has many private good characteristics. However, water is rarely treated as a private good, and as a consequence water is often inefficiently allocated and its use excessive (Dellapenna 1994). The public and private good characteristics and associated externalities differ between different WSS systems (see Table 2.1).

From the perspective of urban service provision, in terms of allocation two issues are of particular importance: the allocation of water between urban and other users in the river basin, and the allocation of water amongst users within the urban area. The allocation of water between urban and other users is especially important in situations where water is scarce and urban areas are growing in terms of both income and population.   In these cases a reallocation of water between different sectors can be expected, often with significant conflict (Hearne and Trava 1997).   As discussed in Chapter 1, excessive consumption by urban users (residential, commercial and industrial) will impact adversely upon competing non-urban uses, such as water used for irrigation of agricultural land. (The converse is, of course, also true.) For instance, in Manila (Angat Dam) urban households using water compete directly with users outside the city. This is not as true in the case of Buenos Aires which draws most of its water from the Rio de la Plata, for which there are no directly competing uses due to the river's massive flow. The over-use of water in urban areas will not only impact upon other water users, including industry and agriculture, but it will also reduce the water available for ecological services such as the maintenance of wetlands and fish populations.

Table 2.1: Public and Private Good Characteristics and Externalities in Water and Sanitation Systems

	Subtractability	Excludability	Externalities[b]
I. Water Supply			
A. Piped			
1. Trunk System	High	High	PH, WD
2. Treatment	High	High	PH
3. Distribution System	Medium	Medium	PH
4. Terminal Equipment			
(a) Common (i.e., handpump)	Medium	Medium	PH PH
(b) Individual (i.e., home faucet)	Medium	High	
B. Village Wells[a]	Medium	Medium	PH
C. Vending (tanker trucks, etc.)	High	High	PH
II. Sanitation			
A. Piped			
1. Main Collection	High	High	PH, EA
2. Treatment	High	High	PH, WR, EA
B. Condominial	Medium	Medium	PH,EA
C. Septic Tanks	Medium	Medium	GC,EA
D. Latrines	High	High	GC

Notes
[a] The degree of subtractability associated with a given well actually depends on the nature of the aquifer from which the well is drawing. High water resource scarcity is assumed. Excludability refers to the tubewell, not the aquifer.
[b] PH = public health, WD = water depletion, WR = wastewater reuse, EA = environmental amenity, GC = groundwater contamination.

Source: Adapted from World Bank (1993b)

Effective conservation and allocation of the use of groundwater is particularly problematic since it is very difficult to ensure excludability of users. Moreover, the volume of groundwater resources is often not known with any degree of certainty. The exact volume of aquifers is difficult to determine, and monitoring of withdrawals is often costly. For this reason, groundwater is often allocated indiscriminately on the basis of firms' and households' willingness and ability to invest in wells. (According to the case studies, this problem is particularly significant in Manila and Mexico City.) Since aquifers span large areas, one withdrawal will have a negative external effect on other users. In any case, where groundwater extraction exceeds recharge, the negative external effects of pumping include the following:

1. The scarcity value of the groundwater stocks is not taken into account, and withdrawn water is not available for other users;
2. The increased cost of pumping from greater depths is imposed upon others;
3. Other users will eventually need to invest in pumps or other sources of water to replace the aquifer upon its depletion;
4. Excess groundwater depletion can even lead to land subsidence (see Mexico City case study) and saltwater intrusion (see Manila case study).

Health and Environmental Externalities

Inadequate WSS provision may result in negative environmental and public health externalities due to unsanitary potable water supplies and inadequate wastewater collection and treatment (see Table 2.1 above). Waterborne diseases include diarrhoea, cholera, and typhoid (see Table 2.2). Significant reductions in morbidity and mortality, especially among children, can be achieved through:

1. Adequate access to safe, potable water;
2. Adequate access to water for washing and cleaning;
3. Proper removal, treatment, and disposal of wastewater and effluent.

Households may well recognise the adverse health effects of these diseases and (if they can afford to do so) adjust their WSS provision accordingly. However, given the communicable nature of many of these diseases, they may not consider the external benefits of their own improved health to the health of the wider community. For instance, a household might choose to use a simple pit latrine which is perfectly sanitary in terms of immediate environmental consequences. However, depending upon soil conditions it may result in externalities by contaminating the groundwater supply of the community. Even if the household itself draws water from this supply, there

Table 2.2: Examples of the Main Water-Related Infections with Estimates of Morbidity and Mortality

Disease	Morbidity	Mortality
1. WATERBORNE (and waterwashed; also foodborne)		
Cholera	More than 300,000	More than 3,000
Diarrhoeal diseases*	700 million or more infected each year	More than 5 million
Enteric fevers (Paratyphoid)	500,000 cases; one million infections (1977-8)	25,000
Infective jaundice (Hep. A)	-	-
Pinworm (Enterobiasis)	-	-
Polio (Poliomyelitis)	204,000 (1990)	25,000
Roundworm (Ascariasis)	800,000 - 1 million cases of disease	20,000
Leptospirosis	-	-
Whipworm (Trichuriasis)	-	-
2. WATER-WASHED Skin and eye infections		
Scabies	-	-
School sores (Impetigo)	-	-
Trachoma	6-9 million people blind	-
Leishmaniasis	12 million infected; 400,000 new infections/year	-
Other	-	-
Relapsing fever	-	-
Typhus (Rickettsial)	-	-
3. WATER-BASED Penetrating Skins		
Bilharzia (Schistosomiasis)	200 million	Over 200,000
Ingested	-	-
Guinea worm (Dracunculiasis)	Over 10 million	-

Notes *This group includes *salmonellosis, shigellosis, campylobacter, E coli, rotavirus, amoebiasis* and *giardiasis*.

Source: Adapted from Hardoy, Satterthwaite and Mitlin (1992)

will still tend to be excess contamination since the household's cost of avoiding this contamination is likely to be greater than the household's expected benefit from better quality groundwater arising from their own efforts. Thus, if on aggregate individuals only account for their personal preferences this service will be under-provided in qualitative terms. Groundwater pollution from inadequate sanitation facilities is a significant problem in Buenos Aires, Manila and elsewhere.

In many cases, dealing effectively with externalities from sanitation requires the provision of collective infrastructure to treat the wastewater which is collected. Pollution of surface waters due to inadequate treatment is of significant concern in all of the case study cities. However, the adverse effects of inadequate treatment will depend upon the characteristics of the receiving waters and their alternative uses. For instance, despite only treating 5 per cent of the wastewater collected, Buenos Aires benefits from the huge assimilative capacity of the Rio de la Plata. In contrast, its two primary tributaries, into which many industries and residential neighbourhoods discharge, are heavily polluted (see Argentina case study).

Urban water users impact upon other water users when they discharge effluent into rivers, estuaries, and the sea. For coastal cities, maritime discharge is often the most cost-effective means of discharge, but can lead to polluted beaches and changes in fish populations. Effluent discharge into rivers reduces the quality of water to downstream users which implies:

1.   Increased treatment costs for municipal or industrial users;
2.   Limitations on the types of crop that can be safely irrigated;
3.   Damages to marine ecosystems;
4.   Reduced amenity value.

Basic Needs and Merit Goods and Services

Access to adequate water supply and sanitation facilities are usually described as basic needs. Amongst other things, this implies that lower-income households will tend to spend a large proportion of their disposable income on water and sanitation (see Table 2.3 for data on expenditure on water for poorer households). Moreover, their expenditure will be proportionately much greater than richer households.[1] For instance, in Mexico, the lowest decile spend just over 5 per cent of their total expenditure on water services, relative to just over 1.5 per cent for the highest decile (see Mexico City case study). In the case of Manila, David and Inocencio (1998) give figures of 8.2 per cent for the lowest income bracket and 0.6 per cent for the highest.[2]

These differences between rich and poor in the proportion of total expenditure allocated to water are not primarily a consequence of differences in consumption levels. Rather, they are mainly due to the inequality in access to public facilities and the relative cost of some alternative sources of water. In fact, non-connection itself can be one of the most important determinants of disposable income for poorer households (see Table 2.4 for a comparison of the cost of vended and piped water in different cities).

Table 2.3: Expenditure on Water for Poor Urban Households

Source	Percentage of Income or Expenditure	Definition	Coverage
Whittington et al. (1991)	18.0%	58% Poorest Households	Onitsha, Nigeria
David and Inocencio (1998)	8.2%	Lowest Bracket (10)	Manila, Philippines
Fass (1993)	3.2% - 10.6%	Spatial Definition	Port-au-Prince, Haiti
Saade (1998)	1.5% - 5%	Lowest - highest decile	Mexico
Cairncross & Kinnear (1992)	16.5% - 55.6%	Squatter Settlements	Khartoum, Sudan

Table 2.4: Ratio of Unit Costs of Water from Vendors and Piped Connections

Source	Ratio	Coverage
Crane (1994)	14:1 - 20:1	Jakarta, Indonesia
Chogull and Chogull (1996)	34:1	Tegucigalpa, Honduras
David and Inocencio (1998)	13:1	Manila, Philippines
Fass (1993)	5.5:1 - 16.5 - 1	Port-au-Prince, Haiti
Whittington et al. (1991)	35:1-300:1	Onitsha, Nigeria
ADB (1993)	62:1	Bandung, Indonesia

Water and sanitation facilities have been characterised not only as basic needs, but as "merit" or "beneficial" goods (Mody 1996, Roth 1987 and Franceys 1997). This implies that society as a whole values private consumption by individuals above and beyond those benefits reflected by personal preferences and external health and environmental benefits. Merit goods have two characteristics. First, they are fundamental to a person's capacity to participate fully in society (see Sen 1983). Dasgupta (1986) calls such goods "positive rights goods". Access to affordable WSS facilities are thought to constitute one such case since they are unarguably fundamental to the realisation of a basic standard of living (see Franceys 1997 and Fass 1993).

Second, there are significant information failures in the provision of water and sanitation facilities. Households do not have access to (or are not able to use) all of the information necessary to make informed choices regarding consumption. Such preference failures are particularly important in demand for WSS due to the complexity of health effects from the consumption of too little water or water of inadequate quality, and from the use of inadequate sanitation facilities.[3]

Thus, even without the existence of health and environmental externalities of the sort described above (e.g. communicable diseases), in the presence of "preference" failures, households may consume too little water or water of inadequate quality. The combined effect of these two characteristics of WSS - the importance of its provision in order to realise a basic standard of living and the potential for information failures related to preferences for its provision - imply that there is a significant demand-side potential for the under-provision (both in terms of quantity and quality) even in the absence of supply-side market failures and health and environmental externalities.

2.3   THE ECONOMIC REGULATOR AND SOCIAL AND ENVIRONMENTAL OBJECTIVES

The characteristics of the WSS sector described above imply that a significant degree of public intervention will always be necessary to ensure that provision is economically efficient, socially equitable and environmentally sustainable. Thus, even if the state does not act as the direct service provider, it still has an important role to play.

In many senses its regulatory functions will not have changed appreciably with PSP. Those implementing agencies with direct responsibility for the realisation of social and environmental objectives may well serve roles which are indistinguishable from those which they fulfilled previously when public authorities had direct responsibility for service provision. As such, the discussion in this section concentrates on the "economic" or "sectoral" regulator since it is its role which changes with increased PSP. Indeed, in most cases the "economic" regulator will only come into existence because of PSP, mainly out of concern for the potential of private sector providers to exercise market power. Thus, through a variety of mechanisms (price regulation, service quality standards, coverage targets, etc.), the regulator seeks to ensure that the service provider does not over-price or under-provide services.

The behaviour of the economic regulator may have very significant environmental and social consequences (albeit often unintentional and indirect). Indeed, in some cases the role it plays may be even more important than that of the agencies with direct responsibility for environmental and social matters. While there is considerable overlap, the effect of economic regulation on the realisation of environmental and social objectives will be discussed in terms of the three market "failures" set out above.

Economic Regulation and Water Scarcity

Given the externalities associated with unsustainable water consumption, it is important that the scarcity value of water be reflected in the costs faced by users. As already mentioned, with respect to urban water supply services, the allocation between urban users and others, and the allocation amongst different groups of urban users are particularly relevant. Implicitly or explicitly, the urban water service provider will be allocated water. This may be done through a number of means including by administrative allocation rules, formalised water rights or permits, or even water charges.[4]

In practice, managers of urban water systems (whether public or private) generally have weak and ad hoc incentives to use water efficiently. For instance, in Manila and Mexico City urban water service providers have not been required to pay the full scarcity value of water from the Angat Dam and aquifers in Mexico's Central Valley, respectively. This is significant since the urban WSS provider's incentives to ration and allocate water use within the urban area is dependent upon the scarcity of water being reflected. In cases where water is allocated freely to an urban service provider and it receives a full priority of the water allocation (so that it does not have to account for the needs of other users), water will not be treated as a scarce resource by the provider. However, if the urban service provider's access to raw surface water supply is constrained in quantitative or financial terms, then it will have an incentive to treat water as a scarce good and to ensure that its customers do so as well.

In their purest form, incentives involve user fees based on consumption. However, this usually requires metering, which can entail high set-up costs. User fees also encourage the repair of leaks and reduction of water losses. If water suppliers pay for raw water, the opportunity cost of water losses, whether through leakages, illegal connections, or non-payment will be higher. This issue is particularly important in Manila where until very recently unaccounted-for water stood at approximately 60 per cent. In Mexico City, it is estimated that 40 per cent of water supply is lost through leakage (see Manila and Mexico City case studies). In Abidjan the figure was 13 per cent in 1987, and the figure was 44 per cent in Buenos Aires in 1992 (see Noll et al. (forthcoming) for a discussion).

In theory, the mechanisms through which the raw water supply is managed sustainably need not change with increased PSP. However, the incentives of the public authorities may well change, with important repercussions for the conservation of water. Under public sector provision, the costs of one public authority (the service provider) are affected by another public authority (the agency responsible for river basin management, for instance). Thus, depending upon political decision-making, water is often supplied to the urban provider without any payment to account for the scarcity of water. If this persists with increased PSP, this represents an appropriation of the resource rent to the service provider and, depending upon pass-through, to urban users. Now that the public authorities are not directly affected by the cost of raw water supply as providers, they may be more likely to charge for the scarcity value of water.

In addition, many of the contracts have specific provisions, which impact upon water conservation. For example, the initial service contracts in Mexico City, which faces significant water supply shortfalls, have been motivated by the desire to manage demand more effectively. Metering of connections, billing of households, and rehabilitation of the distribution network all contribute to the conservation of a scarce resource. These are effective replacements for the rather ineffectual "information" campaigns which were used previously (see Mexico City case study). In many cases, the obligations are even more explicit. For instance, many concession agreements (such as Buenos Aires and Manila) have targets for reducing unaccounted-for water. In some cases, the effects of regulation may be quite indirect. For instance, targets for service expansion can also reduce pressures on water supplies, since unserved households and firms often exploit groundwaters indiscriminately.

Economic Regulation and Environmental and Health Externalities

As with the treatment of the scarcity value of water, the urban water and sanitation service provider is affected by the regulation of environmental and health externalities, both as a user of raw water and as a provider of water and sanitation services. On the one hand, the service provider's costs of production and its ability to meet the public authority's objectives will be affected by the quality of raw water supply. On the other hand, the service provider will be affected by regulations on the quality of potable water and wastewater discharges. The regulations can either be direct (i.e., mandating technologies applied) or indirect (i.e., mandating water and wastewater quality). The latter provides more flexibility and is thus usually more cost-effective. Argentina has recently changed its wastewater quality standards to reflect differences in the assimilative capacity of the receiving waters (see Argentina Case Study).

However, as with measures to efficiently allocate and conserve water resources, there is no *a priori* need for changes to these regulations with increased PSP. Nonetheless, PSP may necessitate the formalisation of regulations to mitigate environmental and health externalities since private firms may have greater incentive to reduce treatment costs than public sector providers had. Moreover, as with efforts to incorporate the scarcity value of water in costs paid by users, the public authorities may be less reluctant to introduce and enforce more stringent regulations when they are no longer the direct managers of water and wastewater services since they will not bear the cost. In some cases, this may even result in an indirect appropriation of the firm's investments (see Noll et al. (forthcoming) for a discussion).

Thus, efforts to mitigate environmental and health externalities are often included directly in the contracts through which PSP is established. Most directly, this arises when specific quality objectives are incorporated into the contracts. For instance, concession agreements often include schedules for upgrading from direct discharge to primary treatment and eventually from primary treatment to secondary treatment (see Buenos Aires case study). Similarly, drinking water quality standards are also often incorporated directly into the contract. In most contracts, failure to meet such standards can result in penalty payments, usually paid for out of a performance bond deposited with the regulator. Attaching such obligations to commercial incentives (and not just public responsibilities) may increase the likelihood of targets being met.

Perhaps more significantly, targets for expansion of the wastewater collection system (or regulated on-site facilities) will also have important effects on local externalities since many on-site sanitation facilities have adverse consequences for groundwater pollution. In most cases the necessary financing requirements are agreed upon with the firm when such targets are established. However, it is still generally too early to tell whether or not such targets are likely to be achieved. Moreover, in many instances, efforts to meet such targets may be undermined by the obligations of the contract in terms of the service specifications provided. Nonetheless, a review of PSP experience does reveal that many concessions have resulted in expansion into areas that were not previously covered.

Economic Regulation and Merit Goods

As with the other market "failures" cited above, state provision is not necessary to ensure that the basic needs of poorer households are met. Rather, the state must serve as a guarantor of a certain level of affordable provision. As a provider, it might do so through increased investment in water and sanitation facilities, a tariff structure which ensures that minimum consumption levels are realised (lifeline or block tariffs), and public subsidies or credit mechanisms for connection fees.

How can it serve as a guarantor of a basic level of provision when it is no longer the direct service provider? Since environmental and social objectives are often (but by no means always) complementary, many of the mechanisms are the same as those set out above. For instance, efforts on the part of the regulator to ensure that private sector providers increase service coverage may give poorer households greater access. Given the past bias against connections for poorer households, this should help them disproportionately. For instance, in Buenos Aires 60 per cent of households without access to formal water service provision are below the poverty line (see Buenos Aires case study). In Manila, the 1998 household survey showed that only about 20 to 25 per cent of respondents in a sample of relatively low income *barangays* (neighbourhoods) have individual MWSS connections (see Manila case study).

In cases where household connections are mandated and poorer households are not able to cover the real cost of the service, straightforward contractual obligations for service expansion will not be sufficient. For instance, in Buenos Aires, households refused to connect to the expanded network offered by the private provider and continued to use informal forms of service provision due to the high cost of connection fees ($1,000). Recent reforms mean that a surcharge on all users (including those who are already connected) has reduced connection costs for new users (see Buenos Aires case study). In other cases (such as Abidjan) connection fees are nil for pipes of a certain size (less than 15 mm in diameter) (see Abidjan case study). This issue is also discussed in Haarmeyer and Mody (1998).

While public subsidies for low-income users may be justified on the basis of social priorities (and external health and environmental benefits), in some cases excessive reliance on public sources of finance may restrict expansion of the WSS system. Although some poorer households will clearly benefit, poorer households in general may, paradoxically, be adversely affected since they have the most to gain from service expansion. There is, therefore, a degree of tension between some contractual obligations which attempt to address distributional concerns in the short-term and long-term distributional objectives (see Chisari et al. 1997 for a discussion). Moreover, given that the costs of service expansion in lower-income neighbourhoods (particularly informal settlements) are likely to be relatively high, this danger may be acute since few concessions prioritise areas for expansion on the basis of socio-economic characteristics.

A number of concessions try to encourage affordability through cross-subsidisation between users, avoiding the need for public subsidies. Cross-subsidy pricing can take a number of forms, rising block rates being the most commonly used. Block rates are used for metered households in Buenos Aires.[5] In Abidjan, the schedule itself is adjusted on the basis of the socio-economic characteristics of the household (Haarmeyer and Mody 1998). (See Bahl and Linn 1992 for a discussion of various alternatives.)

2.4   THE REGULATORY ENVIRONMENT

Contractual obligations can be designed to help ensure that the social and environmental objectives of the public authorities are realised. However, their effectiveness depends on the extent to which the terms of the contract are adhered to by the private sector service provider. This, in turn, depends on the capacity of the regulator, and the relations between it and other public authorities. These two issues will be explored in this section.

Regulatory Failure and Regulatory Capacity

Rather than allowing the state to remove itself from the sector, increased PSP may make the state's role even more crucial. Paradoxically, the need for intervention (rather than investment) may increase with PSP. This places great emphasis on the need to develop an effective regulatory agency. Three potential barriers to the regulator's effectiveness are frequently cited in the literature: technical expertise, rent-seeking and regulatory capture.

The need for technical expertise is self-evident. The job of the state as indirect provider (or regulator) is entirely different from the previous role of the state as direct provider (or manager). While there is likely to be some overlap in skills (i.e., engineers, hydrologists, etc.), in other areas a completely different set of skills will be required. In particular, there is likely to be considerable demand for lawyers to establish and enforce contractual arrangements and economists to interpret the likely implications of these arrangements on the firm and on consumer behaviour. It is significant that in many cases, the bulk of the staff in the regulatory office has come from the displaced public sector provider (see Buenos Aires case study). This may well be necessary for political reasons, but may result in an inappropriate mix of skills within the regulatory office.

Indeed, in almost all cases in which there has been a considerable increase in PSP, a large proportion of the early financing requirements have gone towards attracting and developing staff who are able to regulate the sector effectively. While a certain proportion of the regulator's tasks can be contracted out, there still needs to be a considerable amount of in-house expertise for oversight. In many cases these skills will not be available locally (see Kerf and Smith 1996 for a discussion of the Sub-Saharan African case). However, effective regulation is necessarily political, and requires a good understanding of formal and informal means of decision making. Thus, while foreign expertise may be useful for specific tasks, it is not an appropriate base for an effective regulatory agency.

Rent-seeking arises from situations in which the objectives of the regulator (or individuals within the regulatory agency) are not consistent with the maximisation of social welfare. Previously it had usually been assumed that there was no real difference between "what regulators ought to do and what they actually do" (Helm 1994). However, more recent economic and political analysis (and, indeed, common sense) tends to recognise the potential for strategic behaviour on the part of regulators. This can result in situations wherein the regulator pursues objectives which are not consistent with broader social objectives.

Rent-seeking can take a number of forms. At its most extreme, it might take the form of personal corruption. For instance, the regulator might alter the contractual relationship (or the means by which it is enforced) in a manner which is favourable for the firm in exchange for some reward, such as direct financial payments or job opportunities. This may not be the consequence of corruption *per se*, but rather the outcome of the close relationship which develops between regulators and regulatees. Kerf and Smith (1996) feel that the danger of this in Sub-Saharan Africa is considerable, but that it exists in all countries.[6] Rent-seeking can also take more subtle forms. For instance, the regulator might structure the agreement in a manner which increases its discretionary power, even if this is not in the public's interest. The reward in this case is institutional and professional, rather than personal and financial. Alternatively, if regulators are elected, they may seek to satisfy either the firm or consumers in order to win votes (CEPIS/ECLAC 1997).

Guarding against rent-seeking is difficult. Awarding salaries which are not too far out of line with those being offered by the regulated firms may be necessary. However, in many countries this will not be possible. Salaried employees of the firm compete internationally, but the staff of the regulatory office are often financed from government budgets. Many countries have tried to reduce this asymmetry by "taxing" the provider to fund the regulatory offices (see Argentina case study). Employment policy may also be important. Restrictions on the freedom of regulators to work for private sector providers are often proposed. In the case of Buenos Aires, some of the public sector employees took up jobs with the regulator (ETOSS) and others with the provider (Aguas Argentinas). Clearly, in such circumstances there is a danger of regulators having too great an affinity with the objectives of the provider.

Regulatory capture arises in cases where the regulator is not able (or in some cases, willing) to represent the "public" interest in its dealings with the regulated firm. The most common reason for such a development arises from information asymmetries. For instance, in order to determine an appropriate tariff structure under rate of return, the regulator must have access to detailed cost data in order to determine a fair rate of return (Mody 1996). However, since much of this cost information is internal to the firm, it may be exceedingly difficult to determine. Moreover, the firm has an incentive to make available only that information which inflates costs.

There are two possible ways around this problem. Firstly, there has been a shift toward price cap regulation which does not require such detailed cost data (the Manila and Buenos Aires case studies provide examples of price cap regulation). However, it is widely felt that there has been a general convergence between the two forms of price regulation, with very few "pure" forms of price caps in existence (see Mody 1996 and Foster 1996: Laffont and Tirole (1993) discuss this problem in theoretical terms).[7] Alternatively, more widespread use of benchmarking or yardsticking has been advocated. This allows the regulator to obtain indirectly information which is external to the firm (although not the sector) and overcome some of the asymmetries which are inherent in regulation (see Manila and Mexico City case studies for examples of the use of yardsticking).

The potential for regulatory capture raises a more general concern. International contracts in the sector are dominated by a very small number of firms. The top five firms account for over 50 per cent of all projects involving PSP in developing countries (World Bank 1998). Those firms (the "agents") will learn how to interact with regulatory authorities more successfully with each bid. However, with each contract, the regulator (the "principal") is new. Thus, while the learning curve is steep for both private sector providers and economic regulators, the combined effects of sudden growth in PSP in the sector and the degree of concentration which exists, mean that the "agent" will always be higher up the curve than the "principal". As with shortages in technical skills, public authorities try to overcome this problem by contracting out a number of key regulatory functions, bringing in people with more experience. However, this is unlikely to remove the asymmetry entirely.

Institutional Co-ordination and Common Agency

Responsibility for WSS services may rest at municipal, regional or central government level, or as is the case in many of the larger cities (e.g. in Argentina and Mexico prior to the introduction of PSP) in developing countries, a combination of all three. In such cases, responsibility for different parts of the system (i.e., wastewater treatment, raw water supply, water distribution networks, etc.) will rest with different levels of government. Many countries have undergone sectoral reform (with or without the introduction of PSP) and it has become common to separate the provider institutionally from the other arms of government through the creation of parastatals or state-owned enterprises. By introducing some degree of institutional independence and financial incentive, considerable improvements in service efficiency have been made. For instance, greater reliance on user fees rather than public subsidies and the establishment of performance agreements between managers and public authorities have served to "commercialise" some aspects of service provision.

Table 2.5 reveals how many institutions and agencies are likely to play a role in determining the regulatory environment in which firms operate in the WSS sector. Thus, in addition to the economic regulatory agency, the firm may be subject to forms of regulation from the environment ministry, the health ministry, the rivers and/or coastal waters authority, municipal housing agencies, and land use and planning agencies. The multiplicity of institutions raises the potential for "common agency" problems, whereby a number of regulatory agencies or ministries impact upon the operating conditions of the firm and there is insufficient co-ordination between them.

"Common agency" problems arise when the objectives of different public authorities are contradictory or conflicting and it is not clear which take precedence. In the case of WSS agencies, such problems are particularly significant since there are so many "public" aspects associated with the sector (see Sappington 1996). Thus, if the environmental and health ministries set quality objectives independently of the economic regulator's price regulation, one or other of the objectives is likely to be untenable (see Baron 1985 for a theoretical discussion). For example, in the case of Buenos Aires, the concessionaire (Aguas Argentinas) argued that changes in regulations which prevented the dumping of sludge from wastewater treatment plants at sea were not foreseen when the tariffs were agreed upon. Conversely, in other cases the regulator may set price caps which effectively prevent quality regulators (environmental or health) from introducing the optimal standards.

Any *ex post* adjustments which need to be made have the potential to undermine the credibility of the regulator (Helm 1994). In addition to the short-term costs, this can have long-term consequences. By introducing a considerable degree of uncertainty into the contractual agreement, changes which are external to the contractual relationship (and the regulator's response to them) have the potential to increase risks for the firm. For forms of PSP in which the firm bears responsibility for investment, increased risk can result in under-investment. This problem is particularly acute in the WSS sector where the life of capital is so long. It may also increase the potential for regulatory "chiselling" whereby regulators use the opportunity of such "external" shocks in order to change the conditions of the contract in an arbitrary manner.

Finally, even if by chance price and quality regulation are consistent, there may be excessive administration costs from lack of co-ordination. For instance, the cost and demand data which the regulator requires to regulate prices or rates of return are likely to be valuable inputs in determining appropriate quality standards. As environmental and health ministries and agencies move towards more extensive use of cost-benefit analysis, the administrative costs associated with this lack of co-ordination become more pronounced.

Table 2.5: Market Failures, Regulatory Measures, and Implementing Agencies

Source of Market Failure	Regulatory Measures and Instruments	"Representative" Implementing Institution
Economies of Scale and Natural Monopoly	Firm Entry, Contestable Contracts, Price and Return Regulation	Economic Regulator, Anti-Trust Regulator
Raw Water Conservation and Allocation	Water Markets, Water Allocation Quotas, Withdrawal Fees, User Fees	River Basin Authority
Environmental and Health Externalities	Water Treatment Regulations, Wastewater Treatment Regulations, Increased Access through Subsidies	River Basin Authority, Environmental Agency, Department of Public Health
Merit Goods and Preference Failures	Access Obligations, Lifeline Tariffs, Cross-Subsidies, Public Subsidies, Credit Schemes	Economic Regulator, Housing Authority, Social Security Agency

2.5   CONTRACT DESIGN AND BIDDING PROCEDURES

Although the role of the regulator is clearly important, to a great extent it is secondary, since the contract itself will already be in place. As such, it is important to examine the means by which the contract is designed and awarded (see Kerf et al. 1998). In some senses, the design and awarding of contracts are part of one and the same process since the specifications of the project are often only fully established in the course of choosing the winner. However, some basic parameters are established *ex ante* and these can have important social and environmental implications.

First, the form of PSP (service contract, BOOT, concession, etc.) must be established. This will determine the extent to which efficiency gains and increases in investment levels are likely to be realised. Hence they will also have different social and environmental consequences. For instance, a service contract to meter and bill households (such as in Mexico City), may result in better conservation of scarce water resources than previously but is unlikely to result in increased coverage. Conversely, a more fully-fledged concession (such as in Manila) is likely to result in increased service coverage, and thus may be of greater value to poorer households. However, it must be remembered that more comprehensive forms of PSP place greater burdens on the regulator. Adopting forms of PSP which are not consistent with domestic regulatory capacity will not result in any gains and may even be detrimental (see Kerf and Smith 1996, and Rees 1998).

Second, the scope of the contract in terms of areas of responsibility will be established at the outset. Perhaps the most important issue is whether or not the contract covers both water supply and sanitation. For instance, while the concession granted in Córdoba only covers water supply (with the public authorities retaining responsibility for sanitation), the Buenos Aires concession covers both water supply and sanitation (see Argentinian case study). The social and environmental implications can be significant. If the Córdoba concession had included sanitation, then some of the externalities between water and sanitation would have been internalised, relieving some of the burdens on the regulator.

Third, the spatial coverage of the contract will be established. This can have important social implications. Clearly, it is important that the contract does not allow for "cherry-picking". Since costs of provision in poorer neighbourhoods are often high, but demand relatively low, the danger of this is great. In addition, the successful application of cross-subsidisation is dependent upon there being a certain proportion of users within the system who are able and willing to bear the costs[8] and, as such, will be affected by the spatial limits of the concession area. In Buenos Aires there were discussions about how to define the concession area such that the rich and the poor were not served by different systems (Foster 1996). This would help to avoid problems such as those which arose in Manila where bids for the concession covering some of the poorest areas of the city had higher tariffs than a concession in the other half of the city (see Manila case study).[9] At the opposite extreme, Côte d'Ivoire has a single concession (and tariff schedule) for the entire country.

Once the form, scope and coverage of the contract are established, a procedure must be adopted in order to determine how it is to be awarded (Kerf et al. 1998). There are three basic means of awarding contracts available to public authorities:

- **Competitive bidding/tender**. This can take many forms, but usually involves a series of preliminary "qualification" stages, followed by sealed bids based on pre-established criteria. The Buenos Aires and Manila concessions were awarded on this basis. Mexico City's service contracts also used competitive bidding.

- **Competitive negotiation.** This involves a procedure whereby the public authorities undertake negotiations simultaneously with a number of partners, negotiating over both the nature of the project and the financial terms. The Izmit water treatment BOT in Turkey is one example of a contract awarded on this basis.

- **Direct negotiations.** This usually takes place when a single bidder makes a bid (often unsolicited) to develop a project. The public authorities and the firm then discuss terms and, if satisfactory, a contract is awarded. The Malaysian sewerage concession was awarded on this basis. In some senses, the Abidjan concession was too, once other firms declined to submit bids.

Despite its cost - as much as 5 to 10 per cent of project costs in some cases - competitive bidding is usually advocated by the World Bank and others for a number of reasons (see Crampes and Estache 1997 and Haarmeyer and Mody 1998 for discussions):

- It is more effective at introducing competition into infrastructure projects than other methods of awarding contracts. This is particularly important in water and sanitation since the service provider is usually a monopolist. As such the bidding procedure is the point where the inclusion of competitive pressures is most feasible.

- In addition, competitive bidding procedures are able to reveal information which is of value to public authorities, particularly regulatory authorities. For instance, competitive bidding gives regulators a better indication of the true value of the assets.

- Finally, competitive bidding may make the introduction of PSP in the sector more "legitimate" in the eyes of users and others. This is particularly important if the tariff regime is being reformed, resulting in higher rates for some users.

The competitive bidding procedure usually involves a number of stages. In the first instance, there is a public notification of the intent to invite tenders for a particular project. This is followed by the distribution of bid documents, information on the nature of assets and other relevant factors, and draft contracts to interested parties. Those firms that wish to submit bids then have to go through a pre-qualification stage. This tends to be based on the technical and financial capacity of the firms and does not relate to the specific project to be undertaken. Relevant criteria might include: experience with similar projects elsewhere, the financial strength of the firm, possession of sufficient assets, the quality of service provision in other systems, and other performance criteria such as productivity levels (Kerf et al. 1998). While pre-qualification reduces the number of bids it also serves to reduce the potential for contracts being awarded to firms that are unable to meet the obligations.

For those firms that get beyond the pre-qualification stage there are usually two further steps: the technical proposal and the financial bid. The first represents a detailed description of how the firm intends to meet the objectives of the project as outlined in the tender documents. Depending upon the nature of the project and the contract, the second will be based on factors such as the price paid for existing assets, the lowest proposed user tariff, the lowest construction costs, the lowest public subsidy required or the largest amount of investment. Clearly, the distributional implications of the winning bid will be affected by the criteria chosen. For instance, one which is based upon the price paid to the state will be in the interests of taxpayers while one which is based upon tariff rates will be in the interests of service users (see the Argentinian case study for a discussion of these issues).

In the past these technical and financial proposals were often combined, with the winner determined on the basis of a "weighting" of the two (see Haarmeyer and Mody 1998). More recently, the two proposals have tended to be evaluated consecutively, with some firms being eliminated at the technical proposal stage. In such cases, the financial bid becomes the ultimate criterion by which the contract is awarded, although other factors may be included. In addition, factors which are not strictly financial may also be used as the ultimate determinant of the award. For instance, in the recent contract to develop a water distribution system in Guayaquil, the contract was to be awarded to the firm that agreed to deliver the greatest number of connections. Since investment levels were restricted by the size of the public guarantee, this is merely the quantity-based equivalent to a financial bid.

There are, however, trade-offs in the use of competitive bidding procedures. For example, there may be a trade-off between the transparency of the procedure and efforts to develop innovative means of solving persistent shortcomings in the sector. For instance, it will be difficult to compare bids which have very different technical characteristics (see Haarmeyer and Mody 1998). For this reason, in many cases the tenders involve detailed technical specifications which provide little leeway for alternative solutions. Similarly, there may be trade-offs between flexibility in implementation and the legitimacy of the award procedure. For instance, in many cases it is explicitly prohibited to change the specifications *ex post* after the award has been made since it is feared that this will undermine the legitimacy of the procedure in the eyes of competitors and users (see Haarmeyer and Mody 1998).

Both of these trade-offs are particularly relevant when it comes to the realisation of social and environmental objectives. Since information is likely to be particularly scarce and uncertainty particularly great when it comes to social factors (e.g. willingness and ability of poorer households to pay for different types of service, the costs of provision in areas which have not previously had access to formal services, etc.) and environmental factors (e.g. wastewater quality relative to standards, changes in the stringency of environmental policy, etc.), it may be necessary to forsake some of the attributes of competitive bidding to allow for projects which are flexible and innovative. These issues are particularly pertinent to projects which involve significant amounts of investment such as the concessions in Manila and Buenos Aires. They also affect BOOTs, particularly those which involve water supply distribution and wastewater collection systems. In the case of service contracts (such as in Mexico City), the need for innovation and flexibility is less acute since they involve discrete tasks.

It is the desire to generate "innovative" solutions which has been at least partly behind the use of direct and competitive negotiations in many cases. For instance, in the award of the concession for sewerage in Malaysia this point was made repeatedly, although speed of implementation was certainly a more important factor. However, the negative consequences of such an approach (lack of legitimacy, reduced competitive pressures, less revelation of information) may outweigh the benefits. For this reason it may be preferable to try to introduce the potential for flexibility and implementation into a competitive bidding procedure.

One way of doing so is to have a procedure to choose a "preferred bidder" and then allow for negotiations over the technical and financial characteristics of the project. This was the procedure adopted in Sydney (see Moss 1997). If the means of choosing the "preferred bidder" is transparent, then this may avoid the problem of inflexibility without undermining the legitimacy of the award procedure. Another way of introducing potential for innovation and flexibility is by allowing for a dynamic development of the technical specification of the project through an iterative process. This procedure was recently employed in Guayaquil, where the bidders were given the option of suggesting changes to the specifications. If the changes were accepted by the authorities, then all prospective bidders would be informed.

Innovation and flexibility can also be introduced (without jeopardising transparency and legitimacy) through changes in the contract design. It is important to have clear conditions under which adjustments of the contractual conditions are considered legitimate. For instance, in Manila, despite pleas from the concessionaires, tariffs were not adjusted following water droughts caused by *El Niño* since levels in the Angat Dam Reservoir did not fall by a pre-determined amount.

2.6   CONCLUSION

PSP in urban WSS is likely to increase in importance in developing countries in the coming years. In reducing their role in service provision, governments have not absolved themselves of responsibility for basic social and environmental objectives. For this reason it is vital that the increase in PSP be consistent with the realisation of these objectives. In particular it is important that the following is taken into consideration:

- Water is a scarce resource and the consequences of this scarcity should be reflected in incentives faced by both the service provider and the customer;

- Inadequate provision of WSS can result in significant environmental and health externalities and these should be internalised;

- Potable water and sanitation services can be considered merit goods and as such the public authorities have to serve as guarantors of a basic level of service provision.

These objectives have not changed with PSP, and in many senses the roles of the responsible authorities and their choice of instruments have not changed either. Nonetheless, the presence of PSP does change the incentives of the regulatory authorities and the service providers. It may also result in the application of different (or complementary) mechanisms to achieve these objectives. Some of the main conclusions which have emerged are as follows:

- It is important to ensure that there is a degree of co-ordination between the different agencies that have an impact upon the regulatory environment of the sector. One of the advantages of PSP is that it makes choices and trade-offs between alternative objectives explicit. This should be seen as one of its strengths, rather than a potential source of tension and incoherence. Responsibilities need to be established clearly at the outset.

• In many cases social and environmental objectives may be complementary. For instance, increased coverage in areas which had traditionally used inadequate sanitation facilities will reduce pollution from seepage into groundwaters or runoff into surface waters. However, there can also be important trade-offs between the realisation of environmental and social objectives. For instance, excessively stringent technical specifications for wastewater treatment quality may result in delays in expansion of the system into poorer uncovered areas.

• Thus far, with some exceptions, experiences with PSP in the sector have been concentrated in middle-income countries with considerable administrative capacity. However, in the coming years it is likely that PSP will grow in importance in countries which have even more extreme levels of urban poverty, are faced with greater environmental constraints, and possess less regulatory capacity. The benefits of different forms of PSP need to be weighed against the potential costs if the public authorities are not able to intervene effectively. Thus, a more pertinent issue than whether or not there should be PSP, is the precise form which such PSP should take given existing economic conditions, sectoral objectives and domestic regulatory capacity.

It is important to provide mechanisms that give users a "voice" in the design and implementation of PSP. Most importantly, the technical specifications of the contract should be drawn up in a manner which is consistent with the preferences and budgets of households, and not just the aspirations of public authorities or the experience of foreign private sector providers. In some cases this may result in a degree of service differentiation. It may also result in the devolution of some degree of responsibility for management of certain aspects of service provision to user associations and other groups.

NOTES

1. Bahl and Linn (1992) review a number of country-level studies of water demand in developing countries and find estimated income elasticities ranging from 0.0 to 0.4. This is confirmed by cross-sectional evidence, indicating that the income-elasticity of water consumption is in the region of 0.3 (Anderson and Cavendish 1993). However, it should be emphasised that if the nature of the service provided by the good changes with income then the demand function may exhibit changing elasticities. For instance, higher-income households in which a

significant proportion of water is used for recreation and aesthetic purposes (ie, swimming pools, gardening and car washing) may have highly price-responsive demand. Thus, not surprisingly it has been found that the price elasticity of demand for water differs with income levels, with elasticities being much lower for poorer households (Anderson and Cavendish 1993, Bahl and Linn 1992 and Idelovitch and Ringskog 1997). For related reasons income elasticities may also differ by income level.

2. Relative costs and expenditure on sanitation facilities are more difficult to compare since the variation in service quality is so wide. A number of alternative low-cost on-site facilities provide relatively inexpensive and adequate alternatives to off-site collection systems. For instance, in many cases on-site sanitation facilities will be preferable (in environmental, health and even convenience terms) than some types of off-site sanitation facilities, even though they may be technologically less sophisticated and considerably less costly (see Mara 1996 and Cairncross and Feachem 1993.) However, as population density increases the number of feasible on-site options becomes more constrained.

3. For a discussion of an example in Thailand see Roth (1987). These issues are also discussed in Noll et al. (forthcoming).

4. In much of the western USA rights are traditionally allocated under a first-come-first-served system which gives priority to senior rights which are claimed by creating a diversion for "beneficial use". In Chile water use rights were granted based upon use prior to the 1981 Water Law. This accounts for a distribution of land and water under the Agrarian Reforms of the 1960s and 1970s.

5. Proxies based on type and age of housing are used for unmetered households.

6. Helm (1994) argues that it has been prevalent in the UK.

7. This convergence arises in part from the political infeasibility of retaining regimes that generate profits which are too far out of line with perceived "normal" rates of return. Thus, in order to set the caps, the information requirements may be comparable to explicit rate of return regulation.

8. For instance, if price elasticity of demand is low for high-income users and high for low-income users then it will be easier to cross-subsidise consumption. Unfortunately, this does not appear to be the case in most cities.

9. In some cases it may be possible to have cross-subsidisation across systems. For instance, in South Africa the cities cross-subsidise the townships in some areas (see Nickson 1996).

3. Water and Sanitation in Low Income Neighbourhoods: The Scope for Service Differentiation and Decentralised Management

Nick Johnstone and Libby Wood

3.1 INTRODUCTION

Alternative means that public authorities can use to serve as a guarantor of access for poorer households to adequate and affordable WSS service provision were considered in Chapter 2, including preferential tariff structures, public subsidies and credit mechanisms. However, to one extent or another all of these mechanisms represent a means of making demand more homogeneous, allowing for standardised service provision throughout a service area. In many cases this may not be desirable, and alternative technological and institutional solutions will be required to ensure that the short-term and long-term needs of poorer urban households and neighbourhoods are adequately addressed.

This chapter considers the case for service differentiation and alternative management solutions within a single network service. It examines the impact that PSP may have on the potential for these alternatives to be introduced, as well as the implications they have for the economic regulator. Hand-in-hand with this discussion are the impact that PSP has on regulation of the "non-institutional" sector (discussed in Section 3.2) and the integration of informal service provision with centrally-managed WSS systems. Such integration is consistent with policy developments in the sector towards vertical "unbundling" of different stages of service provision, increased focus on demand-based service provision and decentralised forms of management.

3.2   ALTERNATIVE SOLUTIONS TO WSS PROVISION

Alternative solutions to WSS provision have both technological (service differentiation) and institutional (decentralised management) implications. These two related sets of issues are examined in turn.

Service Differentiation

While in many cases it may be efficient for a single provider to supply a standardised level of service - usually a household water connection and a sewer hook-up - in most urban areas in developing countries this has been the exception and not the rule. Due to decades of inadequate access to public WSS, many poorer households and neighbourhoods have adopted a variety of alternative strategies to ensure that WSS needs are met. Since such measures are often short-term responses arising from a lack of access to formal networks, they do not necessarily represent the most efficient forms of provision. Indeed, they can be expensive and/or inconvenient for the household, and often generate social and environmental externalities (for example, disease, water shortages or pollution as described earlier).

However, many of the unconventional strategies adopted by households, local user associations, small-scale private operators and NGOs can be effective means of providing WSS services at lower cost. For instance, communal wells and standpipes, water vendors and public tanker trucks may be used to supply water, and pit or ventilated improved pit (VIP) latrines, septic tanks and condominial sewerage can be used for sanitation (for examples of different types of service provision see Lyonnaise des Eaux 1998, Mara 1996 and Pickford 1995). Through vertical "unbundling" of the sector, it may be possible to differentiate levels of service provision within a single network. Indeed, there are a number of inter-related reasons why alternative water supply and sanitation solutions may be more appropriate in poorer neighbourhoods.

On the demand side, in a heterogeneous urban area, service differentiation allows for provision to be a better reflection of household preferences and budget constraints. Indeed, in many cases, the costs associated with "over-engineered" standardised levels of service, may exceed the benefits to poorer households, even when external environmental and health benefits are included. Experience in Latin American countries has shown that non-conventional WSS can reduce combined installation and operation and maintenance costs by as much as 75 per cent (World Bank 1998). Lower opportunity costs may mean that systems which require increased time expenditure (for example, collective water points or communal sanitation facilities), but much lower service provision costs, are more efficient for lower-income neighbourhoods.

On the supply side, it may also be possible to allow for the use of local inputs. For instance, for those alternative systems (i.e., condominial sewerage and septic tanks) in which construction and operation and maintenance are more labour-intensive than conventional systems, local labour can be employed in neighbourhoods with relatively low opportunity costs of labour time. Moreover, costs of conventional service provision are often higher in poor neighbourhoods since the physical characteristics of poorer neighbourhoods may increase the costs of the development of traditional infrastructure. Such neighbourhoods are often situated on difficult terrain (for example, landfills, floodplains or highly contaminated areas) and are typically characterised by semi-permanent dwellings and an absence of coherent urban planning (including irregular street layouts and parcels of buildings). This increases both the investment and maintenance costs of conventional supply systems. Often located on the outskirts of cities, or within cities on public or abandoned land, their distance from the main network serves as a further constraint on connection to the central system.

Poor households are frequently found in informal settlements without legal title to land. This makes it more difficult to keep accurate and up-to-date customer records, increasing billing costs for the service provider. Moreover, in such cases, the public authorities are often wary of providing official connections to WSS services for fear that this will "legalise" property rights. Even where the authorities do equip such areas, the inhabitants may move away and settle in new areas because they are unable to afford the increased rents associated with connection, or to acquire rights of possession. Thus, because of the poor security of people and property, and irregular settlement patterns, WSS solutions involving substantial sunk costs which have to be borne by the users are often inappropriate.

Even where service differentiation is not necessarily the most efficient answer in the long term, non-conventional service provision may provide the best solution(s) in the interim period. In this regard, it is important to acknowledge the significant benefits to be gained from alternative solutions - such as public taps - relative to the *status quo*. However, household preferences will change over time with increased income, or even with increased security of tenure. Changing household preferences implies that the technological options chosen by households must allow for upgrading without loss of all sunk investment costs. For instance, VIP toilets can be easily upgraded to pour-flush toilets with sullage, which can in turn be connected to simplified/settled sewerage systems. Similarly, the promotion of neighbourhood resale may be seen as an intermediate progressive solution since it allows for the establishment of less dense networks which will be supplemented at a later stage (see also Lyonnaise des Eaux 1998, Bourne 1984, Mara 1996, Cairncross and Feachem 1993).

Alternative Management Solutions

To some extent, service differentiation generates the potential for the involvement of alternative groups, including local user associations, the informal commercial sector, NGOs and CBOs, in the installation, management and/or operations of WSS systems. Although service differentiation is not a prerequisite to alternative management systems, the latter is often associated with small-scale and/or collective forms of WSS provision in poorer neighbourhoods. Examples include:

- **Delegation of the operation of collective water supply points** to private individuals, or local community or user associations. For example, in Abidjan, retailers pay SODECI for a licence to supply water from public standpoints in peri-urban and slum areas. SODECI controls the resale price.

- **Delegation of the maintenance of sewerage systems** to neighbourhood user associations. For example, in the north-east of Brazil, responsibility for the upkeep and regular maintenance of condominial sewerage systems has been granted to households.

- **Installation of a "master" meter** on the outskirts of the area, with the bill being paid by a local organisation or a third party. The local organisation may also be responsible for maintenance within the area. Such an arrangement has been set up with the assistance of an NGO in Port-au-Prince, Haiti.

- **"Regularisation" of neighbourhood resale** whereby private individuals are entitled to resell water to neighbours but must possess meters so that they are charged for the additional water consumption. Neighbourhood resale was recently authorised in Jakarta, Phnom Penh and Ho Chi Minh.

There are a number of reasons for encouraging alternative forms of management. Perhaps the strongest is the need to recognise and regulate the "non-institutional" sector since a large majority of households - particularly low-income households - in many urban areas rely upon such sources. While many of these enterprises and individuals provide efficient services, in other cases they are exploiting local monopolistic conditions (e.g. control over access to an illegal connection or to a source of raw water) and extracting rents from households.   Water quality may also be inadequate. "Regularising" these services can be of considerable benefit to users.

Integration of alternative forms of management may also assist in developing a feeling of ownership among the target communities. This will tend to foster conditions of security during the execution of the work and subsequent maintenance, as well as adequate conditions of payment for the service and effective integration into the main network. It also fosters an environment of co-operation between the service operator, communities and municipalities to ensure the long-term sustainability of the systems implemented.

Finally, allowing for decentralised management systems may help a large number of service providers to secure a reasonable livelihood. A survey in Guatemala identified over 200 community and privately operated water and sanitation enterprises in Guatemala City alone, while another study in Ethiopia identified 188 such organisations in Addis Ababa (World Bank 1998).   In cases where they are efficient providers, there are potential employment benefits to be realised by integrating, and not displacing, such enterprises.

Some efforts have been made to integrate the development of services which are provided informally with centrally-managed WSS systems (Solo (1998) cites examples from Abidjan, Buenos Aires and Dhaka). However, for the most part they have evolved parallel to central systems in an unco-ordinated manner. Given their importance, it is clear that if basic social and environmental objectives are to be realised, some degree of administrative co-ordination will be necessary. The key policy issue is, therefore, how best to marry the technological efficiencies of centralised development with the political and economic benefits of giving users themselves a key role to play in service provision.

3.3   PSP AND INCENTIVES FOR SERVICE PROVISION IN POORER NEIGHBOURHOODS

There are a number of reasons why PSP may not result in increased service provision in lower-income neighbourhoods. Assuming a standardised level of service provision with PSP there is considerable danger that firms will "cherry-pick" rich neighbourhoods where profitability may be greater. On the one hand, demand for services may be much higher in rich neighbourhoods and supply conditions may be more problematic in poor neighbourhoods, driving up costs. However, in areas where development has been less *ad hoc* this may be compensated for by increased population densities.

In addition, the interests of pre-existing informal service providers may be seen as contradictory to the interests of the formal PSP provider. By formally and explicitly granting monopoly rights to the firm, some forms of formal PSP may actually play an important role in restricting entry into the market to a greater extent than was the case previously under public management. Whereas the public provider may have tended to see informal developments as a welcome means to divest itself of some of its responsibility to provide services, the private provider may see informal systems as a threat to its privileged market position. Thus, there is good reason to expect that encouragement of alternative forms of service provision and institutional structures will be less likely with PSP.

However, there are also reasons to expect that PSP will result in expanded service coverage through service differentiation and alternative management systems. Perhaps most significantly, PSP brings with it a natural tendency towards the increased use of demand-based service provision. Given a contractual obligation to increase the service area, firms will seek the most cost effective means of achieving this (i.e., that which most closely relates to willingness-to-pay for services). For instance, in the case of Buenos Aires, after having recognised that households were unable and unwilling to pay for the services provided, the concessionaire introduced a scheme intended to find feasible alternative solutions to service provision (see Argentina case study).

Although private firms are often better placed to recognise and respond to demand for differentiated services, the extent to which this is realised depends on the form of PSP and the conditions stipulated in the contract. Service and management contracts will not provide incentives for alternative means of service provision since firms are not responsible for investment. Conversely, concessions and BOOTs may do so, depending on the conditions stipulated in the contract and on supply-side conditions. Some contracts have explicitly included obligations for service providers to provide public stand-posts (Abidjan and Manila) and septic tanks (Buenos Aires) in neighbourhoods where households cannot be connected directly due to the costs of, and inability or unwillingness-to-pay for, household connections.

PSP may also provide incentives for the "regularisation" of informal service provision, particularly if the public authorities are not able or willing to ensure the provider's monopoly position in the market. Given the interdependence of alternative means of service provision in a single urban area, the firm will have incentives to ensure the activities of other providers do not impact adversely upon their operations. For instance, insofar as informal service providers are often drawing on the same water source as the formal provider, the firm may find that it is in its interest to establish formal relations with alternative providers. Similarly, if alternative sanitation systems are affecting raw water quality, then the firm will have incentives to "regularise" latrines, septic tanks, and sewerage. In an effort to achieve their own commercial objectives (and meet the conditions of the contract) the firm will have an incentive to reduce such externalities by co-ordinating its activities with the activities of other providers, whether formal or informal.

Moreover, where illegal tapping is prevalent the firm will have a strong incentive to enter into a contractual relationship with the communities represented by the informal sector. For example, in Argentina, most poorer neighbourhoods include users who rely upon water drawn "illegally" (directly or indirectly) from the service network. Such informal operations result in a loss of water from the centralised system and present serious problems in terms of water quality, irregular supply and low pressure. At the beginning of the concession period, almost all - approximately 600,000 - illegal users were incorporated into the Aguas Argentinas centralised system. There are also a number of independent projects which were set up legally prior to PSP and are operated by neighbourhood co-operatives, grassroots groups and local NGOs (see Argentina case study).

However, potential benefits for the firm of integrating informal and formal service provision extend well beyond the internalisation of such externalities. Large companies benefit from economies of scale with regard to the provision of transport, but often face difficulties in metering and receiving payment for water and sanitation services in poor areas. From a private company's perspective, the low consumption rates in poor areas mean that the cost of installing a meter, checking the reading, preparing and sending the bill, and collecting the money for one poor household is often not viable. One way of dealing with the problem of small bills is to move from household to community billing and to leave communities to manage billing internally. Moreover, many firms are reluctant to enter into poor areas with which they are unfamiliar and will therefore benefit from building upon the expertise and networks of existing institutions. Thus, in many cases, sub-contracting, franchising and mini-concessions may be economically efficient.

PSP has often stimulated interest in other institutional forms of service provision. In several cases, the concessionaires have formed partnerships to help provide a different level of service and management structure in low-income neighbourhoods. These relations may be with small-scale private operators, co-operatives, local user associations, NGOs or even neighbourhood groups.

3.4   CONTRACT DESIGN AND REGULATION OF ALTERNATIVE FORMS OF SERVICE PROVISION

While there are clearly potential benefits from PSP, without effective regulation PSP is unlikely to be beneficial to poorer neighbourhoods. Private companies will "cherry-pick" areas where they are able to make the highest profits, and with a standardised level of service provision this will leave many poorer neighbourhoods unserved for considerable periods. However, drawing up and regulating contracts which ensure that the needs of poorer households are met is a difficult task. There may be trade-offs between allowing for sufficient flexibility and at the same time ensuring that the contract is enforceable and in the interests of users.

In most cases the straightforward integration of informal with formal systems of provision is unlikely to be in anybody's interest. In particular, given the economies of scale and technological indivisibilities prevalent in many parts of the sector, it is clear that the feasible set of technological options that the regulator can allow to be integrated into central systems has to be constrained (Kessides 1997).

Moreover, the solutions adopted by NGOs, neighbourhood associations and small-scale private operators are usually short-term responses to the crisis resulting from inadequate access. They are, therefore, unlikely to reflect real preferences or efficient patterns of investment, even from the perspective of the household. Digging private wells, which allow for indiscriminate groundwater abstraction, may result in significant water scarcity and salt water intrusion. Similarly, inappropriate on-site sanitation facilities may result in soil contamination and pollution of surface and groundwater. Thus, perhaps the most difficult task facing the regulator is to ensure that positive aspects of the small-scale operators are preserved, while ensuring that services are provided efficiently and do not generate externalities elsewhere.

This means that the nature of service provision should be addressed at the beginning of the bidding process when clarifying objectives, gathering information and designing the contract. Experience suggests that at present the needs of poorer households tend to be an afterthought and are often poorly reflected in the design of contracts. Even where contractual obligations for the expansion of services into poorer neighbourhoods are set out, in many cases insufficient attention is given to details such as the form provision should take and the way in which it should be managed. It is essential that the preferences of the poor are understood if they are to be provided for efficiently. Involving a variety of organisations familiar with the WSS sector and conditions in poorer neighbourhoods is important since private sector providers may not be best placed (and may not have the incentives) to recognise the potential for suitable non-conventional forms of provision.

The presence of service differentiation and alternative management structures must be reflected in tariff structures as well. It is not in the interests of poorer households to receive lower levels of service if costs of provision are comparable to those received by other households. Indeed, it is startling how often prices for facilities such as public water points exceed tariffs for household connections due to rising block tariffs and free consumption levels (see, for instance, the Manila case study). While this is, of course, counterbalanced by the fact that users do not pay connection fees, it is clear that while there is service differentiation, price regulation regimes need to be designed to better reflect the actual level of service received by the user and the costs incurred by the firm. Similarly, if households and community organisations accept management responsibilities, this should also be reflected in tariffs. If services are integrated, then households should receive compensation for investments undertaken.

Where information deficiencies exist, contract design must be flexible enough to cater for inaccuracies in the assumptions underlying initial contractual agreements. This problem was confronted in Argentina where the initial connection costs associated with expanded household connection coverage proved unaffordable in poorer areas. In this case, alternative payment mechanisms and cross-subsidisation were introduced (see Argentina case study). In other situations, allowing for service differentiation may be desirable, although it may not have been foreseen at the outset. While there is often considerable flexibility to adjust rates, there is less scope for demand-based adjustments in the technical specifications of the contract. This inflexibility can be a significant constraint to expanded service provision. It is significant that the development of alternative forms of service provision in the Buenos Aires case arose despite the regulator's protestations that they were not consistent with the terms of the contract.

Even if the ultimate objective remains a standardised level of service provision, it is important to allow for interim solutions for those areas that will not receive conventional services until the latter part of the concession agreement. Thus, contracts must also be flexible enough to deal with changing socio-economic circumstances and associated changes in household preferences over the duration of the concession. Given the sunk costs involved, it is important that the technical specifications allow for upgrading. The potential for allowing differentiated services in the short-term as a step towards long-term service objectives will be greater where contracts span considerable periods of time. One of the constraints with state provision has been a lack of continuity in policies and programmes and absence of incentives for long-term planning. This has often been aggravated by frequent changes of personnel. For example, in Mexico, directors of utilities stay in their position for an average of less than two years.

In many cases it may not be advisable to grant total monopoly rights to concessionaires if they are not in a position to serve the whole area for a number of years. Pre-existing alternative forms of provision may be required in the interim. This has been implicitly recognised in the case of Manila where the concession does not cover households served by pre-existing alternative licensed forms of provision. In addition, illegal connections have been permitted temporarily, until the providers are in a better position to reach affected households (see Manila case study). It was assumed that the concessionaire would eventually undercut other providers and so there were no explicit contractual obligations to co-ordinate services. However, in many cases the regulator may need to be more pro-active, co-ordinating the relations between different providers.

Consideration should be given to the extent to which the legal and regulatory frameworks enable different actors to become involved in the sector. For example, the regulations concerning water delivery can make it difficult for independent operators to own infrastructure. In fact, the laws relating to PSP are often written with a fixed scheme in mind. In Peru, the recent law defining the management and administration of water purports to open the sector to private participation, but it actually gives municipalities and the local utility companies monopoly rights to exploit and sell water and sanitation services. Public authorities may have to support the opening-up of the water and sanitation sector to small scale entrepreneurs.

In cases where different aspects of service provision are provided by a variety of organisations the regulator may need to play a role in mediating between these groups and establishing clear responsibilities and liabilities. The implications of service differentiation and alternative management in terms of liability and other issues need to be considered. For example, where a master meter is installed at the border of a neighbourhood, the regulator must decide who is responsible for water quality, setting tariffs, maintenance and other concerns within the neighbourhood. As the "supply chain" becomes more extended, involving a variety of different organisations, the job of the regulator becomes even more crucial. The regulator must be able to influence the nature of service provision all the way down the chain, often by indirect means. For instance, it may make the principal concessionaire responsible for regulating the quality and price of services provided by sub-contracted organisations.

4. Private Sector Participation in Water Supply and Sanitation: Realising Social and Environmental Objectives in Buenos Aires

Sergio A. Mazzucchelli, Martín Rodríguez Pardinas and Margarita González Tossi

4.1   INTRODUCTION

During the period 1990 - 1994, a number of Argentine enterprises in sectors such as telecommunications, oil, gas, electricity, water and sanitation, railways, underground transport, and airlines passed into private hands, as part of a broader effort to stabilise the economy. One of the most prominent of these privatisations occurred in 1993, when an international consortium (Aguas Argentinas) won a 30-year concession to provide potable water and sanitation services (WSS) for the City of Buenos Aires and 13 municipalities of Greater Buenos Aires.

The Public Enterprise

Before the concession was granted Obras Sanitarias de la Nación (OSN), a public enterprise, was responsible for urban water supply and sanitation for the Buenos Aires Metropolitan Area (BAMA).[1] OSN also bore responsibility for monitoring and surveillance of direct and indirect pollution of the water sources, the quality of the wastewater discharges and drinking water, and the supervision of its own treatment plants. While in theory OSN was self-regulating, it had no autonomous financing, being funded by transfers from the national treasury. OSN managers reported administratively to the National Ministry of Economy and Public Works, which set rates and planned investment.

There were significant problems with OSN service provision. In 1992 (one year before privatisation) the OSN spent more than 80 per cent of its budget on operations and less than 5 per cent on new investment. This slowed expansion of services (which to all intents and purposes had stopped in Buenos Aires) and system maintenance. Indeed, due to OSN's capital constraints, the majority of new connections were completed and financed through local associations and municipalities.

Given this background, it was felt that if the private sector became involved in the sector (and if the contracts were designed appropriately and regulated effectively) many benefits could be gained. In particular, it was hoped that the involvement of the private sector would broaden the coverage area and increase the quality of services, enhance operating efficiency, provide alternative infrastructure financing mechanisms, and reduce the budget outlay for the public sector.

Environmental and Social Conditions

There had been four main social and environmental consequences arising from the chronic under-investment in the sector:

1. **Highly polluted surface water and groundwater.** An estimated 2.2 million cubic metres per day (million m³/day) of untreated household wastewater and 1.9 million m³/day of industrial effluents flowed across the BAMA into the Rio de la Plata, the city's main source of potable water. In addition, the use of sumps by households without sewer connections (more than 6 million people in Greater Buenos Aires) resulted in groundwater pollution;

2. **Inadequate access to services for disadvantaged families.** At the time the concession was granted, service coverage in Buenos Aires was 70 per cent of the population for water and 58 per cent for sewerage services. However, this coverage was unevenly distributed, with higher levels in the wealthier City of Buenos Aires (CBA). In poorer Greater Buenos Aires (GBA), only 55 per cent of the population of 5.6 million had access to water services, and even fewer (36 per cent) to effluent collection services;

3. **Water and service quality problems for connected households.** Even for those households that were "connected" to the network, the service was of varying quality. Insufficient pressure, excess turbidity and bacterial contamination were recurrent problems;

4. **Inefficient use of water resources**. Due to high levels of waste by users as well as water loss from the distribution system (unaccounted-for water), water use in the region was excessive.

Pressure to overcome these problems was undermined by the fact that government agencies had few of the supervisory authorities which the concessionaires now have (concession authorities, regulatory agencies, and audit agencies). What pressure there was was limited to ad hoc activity in the non-governmental sector, the press, and, to a lesser extent, multilateral credit institutions. Moreover, the absence of clear objectives and explicit contractual goals also hindered these authorities in their supervisory activities. As such, there was a general lack of awareness and institutional support for adequate allocation, management, and rational use of water resources. There were also inadequate levels of information on critical environmental and social issues.

By the beginning of the 1990s, a number of networks were on the brink of collapse. The rehabilitation of old systems and much-needed expansion called for major investments that exceeded the financial capacity of the public enterprise. This situation was exacerbated by an outbreak of cholera in northern Argentina that threatened to spread to the metropolitan area.

The Privatisation Process

Following an international competitive bidding process, in April 1993 the government signed a 30-year Concession Agreement with a consortium led by Lyonnaise des Eaux. Under this agreement, the consortium assumed responsibility for operating, maintaining, and expanding potable water and sewerage services in the concession area. This arrangement enabled the government to retain ownership of OSN's assets and exercise planning authority, while transferring responsibility for service management and investment to the private sector.

Foreign enterprises initially owned 52.6 per cent of the shares in Aguas Argentinas, while Argentine enterprises owned 37.4 per cent, and the remaining 10 per cent was owned by employees through an ownership sharing arrangement. Initially, the foreign enterprises were represented by Suez-Lyonnaise des Eaux (operator of the consortium), Compagnie Générale des Eaux, Aguas de Barcelona, and Anglian Water; and the Argentine enterprises were Sociedad Comercial de la Plata, Meller, and Banco de Galicia y Buenos Aires. In November 1994, The International Finance Corporation (a World Bank affiliate) acquired the equivalent of 5 per cent of the shares in the consortium. The initial and current composition of the consortium and the levels of shareholder participation at the outset and now are provided in Table 4.1.

Table 4.1: Composition of the Aguas Argentinas Consortium

Shareholding companies	1994	1998*
Lyonnaise des Eaux	25.0%	26.85%
Sociedad Comercial del Plata	19.5%	21.62%
Aguas de Barcelona	10.8%	12.72%
Meller	10.2%	5.33%
Programa de Propiedad Participada	10.0%	7.8%
Banco de Galicia	7.7%	8.46%
Compagnie Générale des Eaux	7.6%	7.75%
International Finance Corporation	5.0%	5.12%
Anglian Water	4.2%	4.35%

Notes
*Data at June 1998. In recent months, Sociedad Comercial de la Plata S.A. sold its shares in this and other selected consortia. Most of its shares were acquired by Suez Lyonnaise des Eaux, and smaller proportions by Aguas de Barcelona.

Source: From information provided by Gerencia Departamental de Barrios Carenciados (Low Income Settlements Department), Aguas Argentinas S.A.

Aguas Argentinas was established with $120 million in capital. To ensure the stability of the agreement and to provide adequate operating incentives, the operating company (Lyonnaise des Eaux) is required to maintain at least 25 per cent of the shares in the consortium during the life of the concession. It has a staff of 4,000, more than 95 per cent of whom are from the former OSN.

Under the contract, the firm bears responsibility for all matters concerning the collection, purification, transportation, distribution, and marketing of potable water, as well as the collection, transport, treatment, disposal, and, potentially, marketing of sewerage services, including industrial effluents discharged into the sewerage system. Its responsibilities include technical and commercial operation of the systems, maintenance of the infrastructure, and financing of the investments required to reach specific performance objectives set out in the agreement. Originally, the concession area encompassed the CBA and 13 municipalities of the GBA. The agreement now covers 17 municipalities, as some have been subdivided and the municipality of Quilmes has been annexed.[2] The total population within the concession area is 9 million.

Rather than specifying the required investment, the agreement establishes a series of performance indicators and quality standards for services that must be met within the six consecutive five-year periods. It was estimated that an investment of $4,000 million would be required to reach these goals. OSN staff and fixed assets were transferred to the concession holder. At the end of the concession, all fixed assets are to be returned to the government in adequate operating condition. Recourse to loans will be required as cash flow will cover only 40 per cent of the investment plan for the first ten years.

The Impact of Privatisation

Privatisation has resulted in significant improvements, in particular:

- Increased investment capacity and cash flow, essential for renewed expansion in the sector and for rehabilitation and maintenance of the existing infrastructure;

- Improvement and rehabilitation of potable water and sewerage systems, which were on the brink of collapse;

- The construction of the Northern Wastewater Treatment Plant which will treat the domestic effluents from the municipalities of BAMA's northern zone (700,000 inhabitants) and will discharge into the Reconquista River.

- Access to services for 110,000 people residing in *villas miserias* (squatter settlements) and low income and other disadvantaged neighbourhoods has been provided through the use of various technological and institutional innovations based on participation of the communities, local governments, non-governmental organisations (NGOs), and other grassroots organisations;

- Rates remain below their pre-privatisation levels. Indeed, water is the most inexpensive service in Buenos Aires;

- Improved service quality, in particular, the quality of water distributed. Standards for turbidity, bacteriology, and free chlorine comply fully with international standards;

- A significant reduction in the level of water loss (from 45 per cent at the start of the concession to less than 35 per cent by the fifth year); and

- Significant technical research, including environmental studies on groundwater, environmental evaluations, environmental impact assessments, and social surveys to help plan the concession holder's activities.

However, there have been problems relating to tariffs and environmental issues which have led to the renegotiation of certain parts of the original

Concession Agreement:

1. The progress in expansion has been slower than anticipated since the
    concession holder had to carry out a great deal of rehabilitation work.
    Moreover, high levels of non-payment of the infrastructure (connection)
    charge left the concession in financial deficit which meant that expansion
    was almost called to a halt in 1997.

2. Legislative changes brought into being after the signing of the
    Concession Agreement meant that the environmental standards that the
    concessionaire was to attain were modified.

With the elimination of the infrastructure charge in 1997, the Concession
Agreement had to be renegotiated, bringing expansion efforts to a virtual halt.
A new sanitation plan for the BAMA, which was more ambitious than that
provided in the original agreement, and new tariff rates were agreed.
However, this plan is still awaiting approval.

A number of other problem areas need to be resolved:

*   Many disadvantaged neighbourhoods in zones that are outside of the area
    currently serviced will only receive services in 10, 15, and even 20 years;

*   The present contractual scheme prevents the implementation of non-
    traditional solutions or those designed to expedite the expansion of
    services;

*   In the BAMA, water and sewerage services are provided separately, and
    the latter are still not as prevalent as the former;

*   There is still inadequate understanding of the situation with respect to
    wastewater generation and sanitation;

*   Although the tariff structures are equitable, they provide no incentives
    for expansion to disadvantaged areas or for efficient water use;

*   There are no clear provisions for education about public health and
    hygiene issues;

*   While the concession holder may report illegal industrial dumping, it
    does not have the power or incentives to exercise control in this area.
    Government enforcement is ineffective with regard to control of

industrial discharges, and no solution to this problem is envisaged in the short or medium term.

All of these issues will be explored in subsequent sections, with a focus on the social and environmental implications of private sector participation in the sector.

4.2 THE LEGAL, INSTITUTIONAL AND REGULATORY CONTEXT

The Legal Framework

The concession is governed by National Law 23.696. Performance objectives and service quality standards are explicitly stated in the Concession Agreement (approved by Decree 787/93) and the regulatory framework (Decree 999/92). These decrees establish the basic parameters under which potable water and sanitation services must be supplied, and define the rights and obligations of the concession holder, the Argentine government, and consumers. The regulatory framework was developed concurrently with the transition from public sector to private sector operation. However, new contractual conditions were negotiated which included new obligations, revised tariff rates and investment levels, as well as new social and environmental goals (see Table 4.2).

The concessionaire is also affected by provincial statutes. The relevant laws for the Province of Buenos Aires are Law 11.720 on Special Category Waste Materials, which governs the generation, handling, storage, transportation, treatment, and final disposal of such waste materials, and Law 11.723 which covers the conservation and restoration of the environment and natural resources, and which requires environmental impact assessments to be conducted for works and activities having potential negative impacts on the environment in the provincial territory. The regulation of urban wastewater, water distribution and treatment, are among the activities specifically mentioned in the text.

Finally, the Charter of the Government of the Autonomous City of Buenos Aires guarantees citizens the right to live in a healthy environment and establishes the requirement to conduct environmental impact assessments in connection with both public and private sector development projects and to discuss them in non-binding public consultative hearings.

Table 4.2 The Political and Legal Framework Governing the Concession

Law/Decree/Plan	Object
Law 23.696, August 1989 The State Reform Law	Established the principles for OSN's privatisation programme.
Presidential Decrees 2074/90, 1443/ 91 and 2408/91	Decision to privatise OSN through a concession.
SOP[a] Resolutions 97, 178, 1991	Implementation of the call for bids.
SOPyC Resolutions 53, 186, 1992	Pre-qualification of the bidders and approval of the main bidding document (*pliego*).
SOPyC Resolution 155, 1992	Granted Aguas Argentinas the concession.
Presidential Decree 999, 1992. Approve the regulatory framework for the sector	Set the regulatory framework governing the concession. Established the guidelines for service quality and specified the tariff structure, as well as the rights and obligations of customers, service providers and the Government with respect to the concession.
Presidential Decree 787, Apr. 1993 (Concession Contract)	Approval to grant concession to Aguas Argentinas.
Law 24.051 on Dangerous Substances and its implementing Decree 831/93	Governs the disposal of dangerous substances, including wastewater and sewage sludge. The competent authority (SRNyDS[b]) establishes uses for surface waters and environmental quality standards.
Presidential Decree 149/97	Initiated the renegotiation of the agreement, including adjustments to the infrastructure charge, the environmental management plan for the Matanza-Riachuelo catchment area, as well as plans for water and sewerage services.
Decree 1167/97 of the Executive Branch of the Nation	Approved the Protocol of Agreement signed by the Secretariat of Public Works, Public Services and Works, the Secretariat of Natural Resources and Sustainable Development, and Aguas Argentinas. Allowed for renegotiation of environmental and rates issues. Allowed for the SRNyDS to assess and redefine, if necessary, the applicable wastewater discharge standards for the concession.
Resolution 634, August 1998 SRNyDS	Defined uses for the Rivers Plate and Matanza-Riachuelo.[c] No uses were prescribed for the River Reconquista due to a dispute as to which authority was responsible. Article 3 states discharge permits to the groundwaters have to consider the uses, the environmental quality objectives and the environmental quality guidelines given in Decree 831/93. [d]

Notes

[a] Secretariat of Public Works of the Nation

[b] Secretariat of Natural Resources and Sustainable Development.

[c] The designated use of the coastal fringe of the Rio de la Plata is for water supply for human consumption with conventional treatment, protection of aquatic life and direct contact recreation. River water quality has to meet the appropriate standards required for these uses by the year 2008. In the Matanza-Riachuelo River upper catchment to Ricchieri Motorway the designated use is "direct contact recreation", while in the lower catchment, from the Ricchieri Motorway to the mouth, the designated use is recreation without direct contact. Water quality has to meet the appropriate standards required for these uses by the year 2003.

[d] This change in emphasis from prescribing effluent standards for discharges to receiving water bodies, to setting effluent discharge standards on the basis of the assimilative capability of the groundwater, has required a reconsideration of effluent discharge quality and quantity. This particularly affected the proposals for the South Western and Northern Wastewater Treatment plants.

The activities of the concessionaire are regulated by the following agencies:

- *Ente Tripartito de Obras y Servicios Sanitarios* (Tripartite Sanitation Works and Services Agency - ETOSS). This agency is composed of representatives from the national government, the government of the province of Buenos Aires, and the Municipality of the CBA, and was established to manage the agreement. It is an autonomous regulatory agency which reports to the Secretariat of Public Works of the Ministry of Economy and Public Works and Services and it is responsible for regulating rates, service quality, consumer affairs, and supervision and approval of maintenance and expansion plans for services and investments.

- *Secretaría de Recursos Naturales y Desarrollo Sustentable* (Secretariat for Natural Resources and Sustainable Development - SRNyDS). This is the concession authority. It reports directly to the Office of the President of the Nation, and supervises all environmental issues and related matters. Together with the SOP, it has been actively involved in the renegotiation of the contract with Aguas Argentinas.

- *Instituto Nacional del Agua y el Ambiente* (National Water and Environment Institute - INA). Established in 1996 within SRNyDS, it is responsible for implementing water pollution policy.

- *Secretaría de Obras Públicas* (Public Works Secretariat - SOP). The Secretariat reports to the Ministry of Economy, Public Works, and Services, and is the implementing authority of the Concession Agreement. It participates in all aspects of renegotiation activities.

- *Comité Ejecutor Matanza Riachuelo* (Matanza Riachuelo Executive Committee - CEMR). This is the catchment area authority responsible for implementing the Environmental Management Plan for the Matanza Riachuelo River, and is composed of representatives of the national authorities, the Municipality of the City of Buenos Aires, and the Province of Buenos Aires. This committee is participating in the renegotiation of the Concession Agreement regarding the planning of the expansion of sewerage services and discharges into the river.

The principal change in the regulatory structure brought about by the involvement of the private sector was the creation of ETOSS. Its primary role is to ensure that Aguas Argentinas, as a monopoly provider, does not abuse its privileged position in the market. The board of directors of ETOSS is composed of six members, with equal representation from the national government, the township of Buenos Aires, and the province of Buenos Aires. The agency is divided into six departments and has a work force of 110 and approximately 110 consultants. ETOSS is financed from user contributions (a surcharge of 2.67 per cent is applied to water and sewerage service bills).

Argentina's experience is instructive in that the Buenos Aires concession was the first large-scale experience of private sector participation in the potable water and sanitation sector in Latin America. As has occurred in subsequent cases, most of the staff of the Buenos Aires regulatory agency came from the former public operating body. During the first few years of the concession, ETOSS was required to focus on building its technical and regulatory capacity, and to level the playing field in its relations with the concession holder. During this period, relations between ETOSS and the concession holder were complex and strained.

The concession holder has argued that the internal organisation of ETOSS, which is divided into six departments, creates a problem of compartmentalised responsibilities and sluggish response to complaints. A more substantive issue is the degree of freedom allowed for Aguas Argentinas to meet the goals established in the contract. There is concern that the ability of the concession holder to optimise investment strategies is limited, preventing it from reaching fixed objectives with lower capital investment and higher profits.

Conversely, the availability of accurate, reliable information continues to be a problem for the regulatory agencies. According to ETOSS, the information provided by Aguas Argentinas during the first two years of the concession was "poor, incomplete, and biased". The concession holder, on the other hand, maintains that the regulator requests an excessive amount of information for no clear purpose and does not have adequate analysis and processing capacity. The information transfer process has been improving gradually through the identification of explicit criteria and pre-established periodic automated reporting forms.

Goals and Objectives of the Regulatory Framework and Concession Contract

The general objectives established in the regulatory framework are set out below (Argentine Government 1992a).

- Guarantee maintenance and promote expansion of the system for providing potable water, sewerage, and industrial discharge facilities;

- Establish the regulatory framework that ensures quality and consistency in regulated public services;

- Regulate action and provide adequate protection of the rights, obligations, and powers of users, the concession authority, the concession holder, and the regulatory agency;

- Guarantee the operation of services now provided and those incorporated in the future into one unit, according to the quality and efficiency levels as indicated in the regulatory framework;

- Protect public health, water resources, and the environment.

Table 4.3 Goals of the Buenos Aires Concession

	Start	Year 5	Year 10	Year 15	Year 20	Year 30
Potable water	70%	82	90	95	97	100
Sewerage	58%	66	75	84	90	95
Primary treatment	4%	66	75	83	90	95
Secondary treatment	4%	9	16	83	90	95
Water system renovation	0%	9	12	19	28	45
Sewer renovation	0%	2	3	4	5	5
Unaccounted-for-water	45%	37	34	30	28	25

The initial objectives of the agreement changed as a result of the financial difficulties, and new contractual conditions were renegotiated, which included new environmental and social commitments and goals, and involved rates and investment. Decree 1167/97 approved the Agreement of Protocol, establishing the new terms of the contract. The terms of the Comprehensive Sanitation Plan (*Plan de Saneamiento Integral - PSI*), amending the Sewerage Plan (*Plan Director Cloacal*) for the concession, to be adapted to new environmental and rates objectives, were approved. The established environmental goals refer in particular to the following:

• Provision of service for the entire population of the concession area and to industries with permitted discharge into the sewerage system;

• Interception, transport, treatment, and final discharge of permitted liquid industrial waste and sewerage with a view to ensuring sustainable use of groundwater in the concession area;

• Treatment and disposal of sludge according to current standards.

The need to conduct environmental impact studies and assessments for each project was also specified.

The Buenos Aires Concession Agreement, therefore, may be considered a contract of agreed objectives, in which the concession holder is required to meet established goals, and provide the necessary resources to reach the coverage and service goals. This scheme enables the concession authority (the regulator) to ensure that the concessionaire meets the established objectives, regardless of the economic viability of the concession. Alternatively, with a "means-based contract", investments and resources are specified so that the concession holder's compliance is achieved through the investment and use of the means agreed within the stipulated time frames. "Objectives" are specified, but they do not determine compliance.

These two types of contract are generally used depending on the level of economic and political stability in the country in question. Means-based agreements are used in more unstable environments, as they allow more scope for adaptation. Contracts of agreed objectives tend only to be used in very stable areas, with clear and precise goals. However, in practice, the difference between the two is affected by the nature of the contract management. Even under contracts of agreed objectives, it is always advisable to establish ongoing revision mechanisms to ensure an economic adaptation to changing conditions.

In the case of the Aguas Argentinas Concession Agreement, the objectives were established for a 30-year period in a scheme that does not allow for periodic adjustments. The participants now view this situation as a major impediment to ensuring the economic viability of the concession. In addition, it has a clear impact on the use of non-traditional services to address more critical social and environmental problems or to provide services to disadvantaged populations in peripheral areas that will otherwise be denied access to basic services for decades simply because of their remote location.

It may, therefore, be necessary to move towards a scheme that allows for a combination of objectives and that allows actors other than the concession holder to provide solutions.[3] While it would be the concession holder's responsibility to oversee the input of other actors, it might be possible for those other actors to expedite solutions to critical environmental and social issues. An adequate mechanism would need to be established to give the concession holder appropriate incentives to operate efficiently in reaching the agreed goals within a stable political and legal framework.

4.3   THE ENVIRONMENTAL AND DEMOGRAPHIC CONTEXT

Environmental and Geographic Issues

The BAMA covers an area of approximately 388,000 hectares. Originally a number of tributaries traversed the city, flowing into the Rio de la Plata. The particular features of a plains region can be found in all of these waterways: a short, shallow course with low permanent flow volume, irregular water courses, broad flood valleys, and drainage and runoff problems resulting from the absence of slope, particularly in the lower stretches. Lack of foresight, inadequate policies specifying territorial settlement guidelines, insufficient investment in the infrastructure relative to demographic growth, and inadequate management of the major rivers and streams running through the area have led to serious environmental problems, ranging from polluted waterways to floods in the vast lowlands, many of which are densely populated.

The Rio de la Plata, which is the BAMA's natural eastern boundary is adversely affected by two rivers (the Matanza-Riachuelo and the Reconquista) and numerous streams, carrying highly pollutant industrial effluents, rainfall runoff, and illegal household and industrial discharges. The Matanza-Riachuelo River runs through the BAMA's most dense, industrialised areas to the south and west. This river is highly polluted by the discharge of industrial effluents. The catchment area of the Reconquista river covers a number of townships in the northern BAMA section, some of which have a longstanding industrial tradition and which use the river as a receptacle for their contaminated waste. The banks of both the Matanza-Riachuelo and Reconquista rivers are largely settled by disadvantaged populations, further contributing to pollution levels due to inadequate access to sanitation facilities. Most of the remaining waterways traversing the CBA and GBA are slow-running and are carried entirely or almost entirely through pipes and channels.

In total, an estimated 2.5 million m^3/day of untreated wastewater and 1.9 million m^3/day of industrial waste flow through the BAMA and drain into the Rio de la Plata. Of the principal pollutants, residential wastewater is estimated to represent approximately half of the biological oxygen demand and two thirds of total suspended solids (World Bank 1995). In addition, the sewer effluent collection system, with more than 7,000 kilometres of conduits, transports approximately 2.2 million m^3/day of effluent. The wastewater has always been discharged into the Rio de la Plata at the township of Berazategui, through an outlet that transports the final discharge approximately 2.5 kilometres from the river's edge, downstream from the two principal water intakes. These effluents are not treated prior to discharge. The only liquid sewerage treatment plant is the South Western plant located in the township of La Matanza, which only processes a fraction (5 per cent) of the effluent. This plant discharges the treated effluents at Riachuelo. The Northern purification plant is at the completion stage, and will treat liquids from the townships of northern GBA (700,000 people).

Direct and indirect discharges of untreated wastewater into the Rio de la Plata are significant, since the river is the BAMA's largest source of water for consumption (92 per cent), with the remainder of the water supply extracted from groundwater sources. The water (approximately 20,000 m^3/second) is provided through two major intakes (Palermo and Bernal). Despite the pollution present in the coastal area, the cost of water treatment for consumption is low. Favourable oxygenation and dilution levels mean that conventional treatment for consumption can be used. The water is treated at two purification plants (San Martín and General Belgrano), supervised by the central laboratory of Aguas Argentinas, one of the largest facilities of its type in the region. The San Martín plant is the largest of its kind in the world. The combined output of the plants is 5 million m^3/day.

Use of the Rio de la Plata's water for consumption does not compete with other uses such as irrigation or power generation. While the banks of the river are not fit for bathing or other sports (direct contact or otherwise), sports activities do take place, with the attendant risks. Where fishing is concerned, the quality of the catch is low, as heartier and less valuable species tend to predominate, owing to the prevailing river conditions. The deterioration in the immediate coastal area does not appreciably affect maritime activities, as boats quickly pass the coastal fringe en route to high-quality areas (more than 500 metres from the coastline).

Groundwater resources in the BAMA are composed of the Epipuelche, Puelche, and Hipopuelche aquifers, listed in increasing order of depth. Of these aquifers, only Puelche is fit for human consumption and has been the largest groundwater source for domestic and industrial consumption.[4]

However, this aquifer has deteriorated as a result of both over-consumption and pollution, with cones of depression and high nitrate concentrations at different points. In some sections of Buenos Aires, the nitrate concentration in the groundwater exceeds the limits established by the World Health Organisation (WHO) of 10 mg/l. Further, the arsenic concentration in some wells exceeds tolerable limits of 0.05 mg/l. Mercury and chromium contamination was also detected.

The main source of groundwater pollution in the BAMA is the proliferation of septic tanks and sumps in the areas without sewerage services. Industrial wastes are also often discharged into septic tanks and cesspools designed in part to discharge into the Epipuelche aquifer. Since the commencement of the concession, Aguas Argentinas has suspended the operation of more than 100 wells owing to high nitrate concentration. In 1994, four wells were suspended as the result of the presence of chromium.

Social and Demographic Factors

The BAMA can be divided into three sectors, which until 1991 had the following features[5]:

- **Central Buenos Aires** (former Federal Capital), which covers 7.2 per cent of the area and 27.9 per cent of the population. This sector included the highest population concentration, the lowest level of demographic growth, and the greatest coverage in terms of services, infrastructure, and facilities.

- **The first ring** corresponds to the areas bordering the CBA, covering 26.6 per cent of the area and 46 per cent of the population. Increased density in this area can be attributed to higher levels of investment in infrastructure and services.

- **The second ring** has the highest population growth rate in recent decades, is most recently urbanised, and has major health and social infrastructure deficiencies. Many of its population's basic needs are not satisfied. This area still includes unoccupied government and private land. It covers 66.2 per cent of the area, but only 24 per cent of the population.

The spatial pattern of economic development in the BAMA is not homogeneous. The driving force of the region's economy comes from an industrial axis covering the coastal areas of the Rio de la Plata. Moving away from the coast, tertiary and secondary activities (less concentrated in other municipalities of GBA) become predominant.

This uneven pattern of development has led to an unusual distribution of land uses. The upper municipalities of the main river valleys (Luján, Reconquista, and Matanza), have better sanitary conditions and are used for recreational activities or non-permanent residential purposes. Conversely, in the middle and lower sections of these rivers - which are prone to periodic flooding, and have inadequate sanitary conditions - industrial areas and residential areas, the latter occupied primarily by disadvantaged groups, exist side by side. The elevated central areas house higher-income sectors.

Comparing this with system coverage in Tables 4.4 and 4.5 reveals that it is the poorer households that are most affected by under-investment in WSS services in Buenos Aires. This index can also be compared with service coverage levels. Figures 4.1 and 4.2 show the percentage of households of different socio-economic categories with water and sewerage connections. Not surprisingly, the higher socio-economic categories have the highest coverage levels, both for water supply and for sanitation.

Figure 4.1: Households with Water Connections by Socio-economic Classification

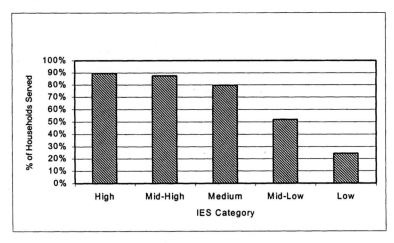

Source: From information provided by Gerencia Departamental de Barrios Carenciados (Low Income Settlements Department), Aguas Argentinas S.A.

Figure 4.2: Households with Sewer Connections by Socio-economic Classification

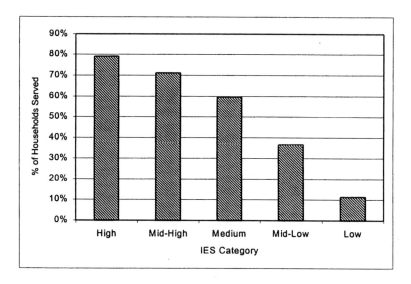

Source: From information provided by Gerencia Departamental de Barrios Carenciados (Low Income Settlements Department), Aguas Argentinas S.A.

Table 4.4: Population with Water and Sewerage Services at the Commencement of the Concession

Area	Total population 1991[a]	Population growth Index 80/91[b]	Population with running water service		Population with sewerage service		Agency providing the service
			Population[c]	%[d]	Population[c]	%[f]	
City of Buenos Aires	2,960,976	1.3	2,922,000	98.77	2,922,400	98.6	Obras Sanitarias de la Nación (OSN) (Currently provided by Aguas Argentinas)
Vicente López	289,005	-0.7	282,580	97.7	275,340	95.2	
San Isidro	298,540	3.2	265,240	88.8	104,280	35.0	OSN
San Fernando	142,925	7.0	70,100	49.0	34,010	23.8	Currently provided by Aguas Argentinas
Tigre	255,041	23.6	54,490	21.3	20,720	8.1	
San Martín	403,515	4.6	229,590	56.8	162,940	40.3	
Tres de Febrero	349,221	1.1	230,100	65.8	159,200	45.5	OSN
Morón	641,416	7.2	189,910	29.6	149,190	23.2	West Zone
La Matanza[g]	1,120,225	18.0	373,980	33.3	399,910	35.7	(Currently provided by Aguas Argentinas)
Avellaneda	338,581	1.3	276,680	81.7	109,950	32.4	
Lanús	465,454	-0.3	354,650	76.0	72,620	15.6	OSN
Lomas de Zamora	572,318	12.2	315,960	55.0	166,920	29.1	Southern Zone

Alte. Brown	443,251	33.5	51,570	11.6	35,100	7.9	(Currently provided by Aguas Argentinas)
E. Echeverría	273,779	44.9	27,840	10.0	21,360	7.8	
Quilmes	509,449	14.1	377,345	74.0	172,212	34.0	Municipality (Currently provided by Aguas Argentinas)
Berazategui	244,796	21.3	154,000	63.0	34,271	14.0	[h]
Fcio. Varela	255,462	47.3	20,463	8.0	17,882	7.0	General Administn
Gral. Sarmiento	650,285	29.3	29,262	4.5	26,011	4.0	OSN
Merlo	385,821	31.9	16,976	4.4	14,455	3.7	Prov. de Buenos Aires
Moreno	287,295	47.8	14,365	5.0	11,492	4.0	(AGOSBA)
Subtotal AGOSBA	1,578,863	26.8	81,066	5.1	69,840	4.4	AGOSBA
Total Municiplties Greater Buenos Aires	7,926,379	15.8	3,335,101	42.0	1,988,863	25.0	OSN-AGOSBA Townships[i]
Total BAMA	10,887,355	13.4	6,257,901	57.0	4,911,263	45.0	19 townships and Federal Capital

Notes

a and b Provisional figures from the 1991 national census (INDEC).

c, d, e and f Estimated according to number of OSN connections 1990, AGOSBA and townships, according to SRH-INDEC, 1985.

g Includes the Federal District Ezeiza which corresponds administratively to OSN's western zone.

h OSN provides water which is distributed by the township.

i It is estimated that illegal water service users in the OSN's territory amount to 189,910 persons and that disconnected OSN and AGOSBA services account for 244,140 persons.

Source: Pescuma and Guaresti (1991)

Table 4.5: Population Without Water Service Coverage at the Commencement of the Concession

Area	Total population without service	Unserviced disadvantaged population		Population without running water service				Deficient discharge	
				Well		Other System			
Municipalities of Greater Buenos Aires	Pop[1]	Pop[2]	%[3]	Pop[4]	%[5]	Pop[6]	%[7]	Pop[8]	%

BAMA townships within the Aguas Argentinas S.A. concession

Vicente López	6,425	706	11	3,726	58	2,699	42	8,940	3
San Isidro	33,300	5,661	17	27,972	84	5,328	16	14,975	5
San Fernando	72,825	29,858	41	48,792	67	24,033	33	28,585	20
Tigre	200,551	72,198	36	160,440	80	40,111	20	58,659	23
San Martín	173,925	46,959	27	144,358	83	29,567	17	32,281	8
Tres de Febrero	119,121	25,015	21	95,297	80	23,824	20	17,461	5
Morón	451,506	94,816	21	397,325	88	54,181	12	51,313	8
La Matanza	720,315	280,922	39	597,861	83	No data	-	No data	-
Avellaneda	61,901	30,332	49	55,130	89	6,771	11	20,314	6
Lanús	110,804	38,781	35	58,726	53	52,677	47	51,120	11
Lomas de Zamora	256,358	87,161	34	182,014	71	74,344	29	108,740	19
Alte. Brown	391,681	137,088	35	270,259	69	121,422	31	97,515	22
E. Echeverría	245,939	90,997	37	210,809	85	34,584	15	68,445	25
Quilmes	132,104	66,052	50	80,583	61	51,521	39	96,795	19

BAMA townships operated by other agencies

Berazategui	90,792	40,858	45	31,823	35	58,973	65	46,511	19
Fcio. Varela	234,999	112,799	48	183,932	78	51,067	22	94,520	37
Gral. Sarmiento	621,023	229,778	37	540,290	87	80,733	13	162,571	25
Merlo	368,845	140,161	38	317,207	86	51,638	14	96,455	25
Moreno	272,930	120,089	44	229,261	84	43,669	16	91,934	32
Subtotal AGOSBA	**1,497,797**	**602,827**	**40**	**1,270,690**	**85**	**227,107**	**15**	**445,480**	**28**
Subtotal Great Buenos Aires	**4,591,279**	**1,650,231**	**36**	**3,635,805**	**79**	**955,473**	**21**	**1,348,726**	**17**

Notes
1, 2, 4, 6 and 8 INDEC, 1985.
3, 5 and 7 Percentage of unserviced total.
6 Includes wells, service reservoirs, and water carriers.
8 Inadequate disposal: opulation with unconnected toilet (squatting latrine) and without a toilet
   of any kind.
9 Percentage of total population of the area.

Source: adapted from Pescuma and Guaresti (1991)

4.4   THE TARIFF AND RATES STRUCTURE

In this section the rates structure is analysed, looking at both efficiency and equity implications, through a comparison of the Buenos Aires rate structure with that of Córdoba, which has recently granted a similar concession for water (but not sanitation) services. This helps to illustrate some of the potential differences in terms of efficiency and equity which arise from apparently small differences in tariff structure.

General Principles

The guiding principles behind the rate structures in place in the two cities are similar, as shown in Table 4.6.

Table 4.6: Rates Principles

Córdoba	Buenos Aires
Will promote rational and efficient use of services provided to users and resources applied for that purpose.	Will promote rational and efficient use of services and resources involved for that purpose.
Will enable consistent equilibrium between supply and demand of services and ongoing improvement and expansion in these areas.	Enable consistent equilibrium of supply and demand for services to be attained.
Will support health and economic objectives in social matters related to the service.	Support health and social objectives related directly to the service.
Should reflect the economic cost of providing services with efficiency as a criterion, incorporate emerging costs of improvement and expansion plans, and, when applicable, reasonable profit margins for the service providers.	Prices and rates will tend to reflect the economic cost of providing services, including the concession holder's profit margin and incorporating the emerging costs of approved expansion plans. Enable cross-subsidies to be applied.

Both concessions apply the rates system that was used by the OSN since the mid-1970s. At the beginning of the 1980s, the system was decentralised, and service in the province of Córdoba was transferred to the provincial body, Obras de la Provincia, which maintained the rates system, and only changed some parameters to adapt to the new conditions. Both systems have a mix of metered and unmetered connections. Rates for unmetered connections are charged periodically (bi-monthly in Buenos Aires and monthly in Córdoba) and based on property taxes, adjusted by a variety of coefficients. Rates for metered households have a fixed charge derived from the bi-monthly charge for unmetered households, and a variable charge per unit of water consumed.

Concessions were awarded on the basis of the greatest discount on the basic rates. Consortia were requested to submit sealed bids, expressed as reductions in the coefficients (AC) used to adjust rates across time.

Table 4.7: Criteria for Award of Concessions

Aguas Argentinas (Buenos Aires)	Aguas Cordobesas (Córdoba)
Lower AC was exclusive criterion for awarding the concession.	Awarded on basis of lower AC, size of royalty payment, and absorption of public employees.
Original 0.731 changed to 0.830 to reflect increased investments in the year the concession began.	0.918, weighted by zone*

Notes
* The AC is differentiated by zones (0.8 for poorer zones and 1.5 for richer zones)

However, in the case of Córdoba the decision criteria also included the royalty payment by the concession, absorption of public employees, and wage levels. The awards for lower rates and higher royalties have similar features with respect to the selection of the most efficient bidder from the productive standpoint. The difference lies in how the efficiency gains are distributed. Lower rates pass the gains of productive efficiency directly on to users, and comply in principle with allocative efficiency by reflecting production costs in rates. The higher payment gives the concession authority (the government) an income that increases the rate established in relation to the cost of providing the service.

The use of multiple criteria requires weightings for each element in the selection process. The relative weights of each element in the case of the Córdoba concession are provided below:

Rates discount	60%
Royalty payment	20%
Employment:	20%
Wages	10%
No. of Employees	10%

In all cases, bids were considered in relative terms (as quotients between the best individual value and the one under consideration).

Interestingly, the successful bidder for the Córdoba concession proposed a royalty payment of $9.9 million. If we consider that the concession's annual billing is approximately $50 million, we estimate that an award based exclusively on the rate would have yielded a value some 20 per cent lower.[6] Thus, the distributional implications of using a royalty payment to award the concession is significant, resulting in a cross-subsidy from users to taxpayers.

Rates for Unmetered Systems

While the Córdoba concession has a single rate for all classes of unmetered user, in Buenos Aires rates are differentiated by type of consumer with a 100 per cent difference between residential and non-residential customers. Accordingly, a substantial cross subsidy is introduced from commercial and industrial consumers to residential consumers.

Table 4.8: Residential and Non-Residential Rates

Buenos Aires		Córdoba
TG(CAT):		
CAT=Res	0.0279 $/m^2	0.0200 $/m^2
CAT=NonRes	0.0558 $/m^2	

The problem with this type of subsidy is that, by not distinguishing between residential customers, all users receive the subsidy regardless of their real payment capacity. Accordingly, small commercial establishments in outlying disadvantaged neighbourhoods subsidise consumption by large dwellings in higher-priced neighbourhoods in the city. In addition to the adjustment coefficient, a variety of coefficients are used to adjust the basic rate, based upon the following characteristics:

- Covered area of the building (CS);
- Uncovered area of the building (LA, with a weighting factor of 0.1 relative to CS);
- Zone coefficient (Z);

- Building age and type coefficient (C).

The choice of proxies reflects the twin objectives of equity and efficiency. While increased efficiency is the primary objective when restructuring public service provision, for water services equity is an important additional component in the formulation of rates, as it is an essential element for human life.

Of the four variables used in the rates formula, building area and land area are designed to reflect consumption, with the assumption that a larger area is associated with higher levels of water consumption. The aim is to meet the allocative efficiency objective as the rate is related to the cost of providing the service (higher consumption levels, reflected by a larger area, corresponding to a higher rate). Note, however, that with unmetered consumption such "proxies" are not able to provide marginal incentives for more efficient water consumption at the level of the user.

The third and fourth variables are designed to reflect ability to pay, promoting a certain degree of equity in payment of rates. Given the considerable externalities associated with potable water consumption, it is desirable and, in many cases, efficient to subsidise poor users to protect the health of the population. Using the zone coefficients and the construction type, equity objectives can be incorporated. The building coefficient has two variations depending on the type of construction (divided into four categories), and the year of construction. Accordingly, the more modern the construction, the higher the water rate, to reflect a presumed greater payment capacity associated with owners of newer housing.

Building coefficient

The building coefficients are different in the two schemes, with the Buenos Aires concession dividing the coefficient into more building categories (six, as compared with four for Córdoba). Of these categories, the four intermediate ones correspond to those used in Córdoba.

In addition, while the values observed for the coefficients coincide perfectly until 1980, as we observe in Figure 4.3, thereafter there is a differentiation between the two enterprises both in terms of level and rate of variation of the coefficients. The period for which the building coefficient values coincide corresponds to the period during which OSN was responsible for all services in Argentina, after which responsibility was transferred to the provinces.

Table 4.9: Building Coefficients

	Buenos Aires	**Córdoba**
Type of construction		A (age, CAT, construction quality)
Age	Coincide until 1980 (transfer to the provinces)	
	Lower values during the period 1980-1997 Defined for the entire concession	Higher values during the period 1980-1997 Defined only until 1997
Construction quality	Six categories	Four categories coincide with the four intermediate categories of AA

Figure 4.3: Building Coefficients for Aguas Cordobesas and Aguas Argentinas

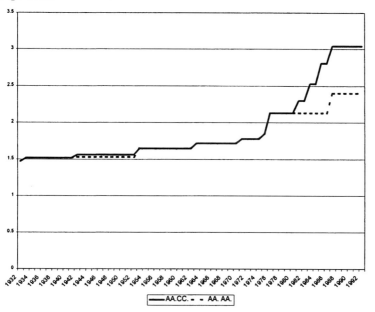

One of the reasons for this is the difference in the population density in the two concessions. The Aguas Argentinas concession covers an area of 774 square kilometres with 2.5 million customers and a total population of more than 10 million, i.e., a density of 3,262 customers per square kilometre. The Aguas Cordobesas concession covers an area of 270 square kilometres with

350,000 customers and a population of one million, i.e., a density of 1,296 customers per square kilometre.[7] Similarly, while the BAMA has 180 connections per kilometre of system, the ratio in Córdoba is only 140 connections per kilometre (see Table 4.10). These differences are significant given the importance of network fixed costs in the total costs of providing the service. Clear density economies are observed. The greater the density of the service area, the lower the unit cost of supply, as the fixed costs of the common system can be divided among more users.

Table 4.10: Comparative Service Density in 1997

	Aguas Argentinas	Aguas Cordobesas
Serviced area (km²)	774	270
Customers (000)	2,525	350
Population (000)	10,000	1,000
Network extension (km)	13,975	2,494

This difference in density is the principal economic justification for the different developments in rates in Córdoba and Buenos Aires. While the overall system was in the hands of the same enterprise, a regional equity criterion was applied, and, accordingly, a single rate existed. This resulted in subsidies from the more densely populated to the less densely populated regions. A second factor that explains the differences in rates is the differing investment requirements of the two systems at the time of the transfer. While coverage in the BAMA increased from 65 per cent in 1980 to 70 per cent in 1991, coverage in Córdoba rose by almost three times that level, from 71 per cent to 84 per cent (see Table 4.11).

Table 4.11: Water Service Coverage by Province

	Coverage	
	1980	1991
Area		
BAMA	65%	70%
Rest, GBA	32%	35%
Rest, Prov. of Bs.As.	63%	67%
Córdoba	71%	84%
Santa Fe	62%	74%
Mendoza	86%	90%

Source: Adapted from IDB (1996)

After the service was transferred to the provinces, each enterprise was expected to be self-financing, and thus disparities appeared in the rates they charged to reflect existing cost differentials. One of these disparities relates to differing building index trends. In the case of Córdoba, the increase is greater than it is for OSN, which provided the service in the CBA and Greater Buenos Aires. From the time of transfer to the provinces until the privatisation of the enterprises, the index for Córdoba was more frequently adjusted than had been the case under the OSN, reflecting an intention to provide true financing for higher costs associated with the special characteristics of Córdoba's case.

The greater increase observed in the building index in Córdoba reflects differences in costs and the fact that the relative index structure did not change, keeping the ratio between the different groups unchanged. More frequent adjustment of indices and higher relative growth levels are associated with differences in expansion costs, rather than differences in rates applied by the enterprises. The changes may also reflect a certain level of political expedience, as the coefficients for new housing were raised before a general rate adjustment or complete revision of the coefficients. Raising the coefficient only on new housing does not affect users already receiving the service, thus costs are concentrated on new construction.

Interrelationship between the building coefficient and type of construction

In the case of water services, the same trends in the age coefficient over time are observed for all types of buildings defined. For instance, a house built in 1998 pays twice as much in Córdoba and 1.8 times more in Buenos Aires than a house built prior to 1932, regardless of whether it is a luxury or a slum dwelling. It has been estimated that overall cross-subsidies in 1995 were as high as 7:1 for properties of similar size but different characteristics (see Noll et al., forthcoming).

Table 4.12: Building Coefficient Ratio by Age of Construction (1998 & 1930)

	Córdoba		Buenos Aires
	Water	Sewerage	
Luxury			1.80
Moderate	2.07	1.65	1.80
Good	2.07	1.65	1.79
Moderately Low-cost	2.06	3.28	1.79
Low-cost	2.07	4.41	1.80
Very low cost			1.81

However, in the rates system for sewerage in Córdoba, the ratio varies substantially among different types of housing. While higher-quality construction has a coefficient of 1.65 between the most recent construction and constructions of more than 50 years old, the coefficient for low-cost housing is 4.41. To the extent that housing age is used as a proxy variable for payment capacity, differentiation by type of housing is assumed to be necessary for proper identification. Accordingly, it seems reasonable to assume that age has a greater impact on housing value for lower construction quality than for a luxury unit. The opposite is clearly also true – the lower the construction quality, the greater the effect of time on the value of construction. Age therefore becomes an important factor in identifying payment capacity in this case.

Minimum charges

In order to cross-subsidise households, both concessions apply minimum charges. However, they do so by rather different means. While in Córdoba, minimum charges are a direct function of assumed payment capacity (reflected by the zone coefficient Z), in Greater Buenos Aires, a differentiation by type of consumer is applied. As a result of this differentiation, the Aguas Argentinas scheme establishes a cross-subsidy from non-residential (NR) to residential (R) users, and to a lesser extent, to undeveloped land (U).[8]

Table 4.13: Minimum Periodic Charges

Buenos Aires (bi-monthly)	Córdoba (monthly)
MIN(CAT)	MIN(Z)
CAT=R -> MIN = AC * $4.00	Z=0.8 -> MIN= $4.32
CAT=NR -> MIN = AC * $8.00	Z>0.8 -> MIN= $8.11
CAT=U -> MIN = AC * $3.00	

The reformulation of the infrastructure charge which arose as part of the renegotiation of the Buenos Aires concession has distributional implications. If new connections are to be financed by a fixed amount ($2) on all bills rather than by applying a proportional charge to existing bills, this will reduce the cross subsidy from non-residential to residential users and to a more-than-proportional increase (by up to 66 per cent) in minimum bills. [9]

Zone coefficient

The range of the zone coefficient differs in the two concessions. Although the theoretical range of Aguas Argentinas exceeds 400 per cent (0.8-3.5), the actual rate is limited to a range of 270 per cent. Córdoba's range is 216 per cent.

Table 4.14: Range of Zone Coefficients

Buenos Aires	Cordoba
Theoretical range: 0.8 - 3.5 Effective range: 1.3 - 3.5	Effective range: 0.8 - 1.73

Depending upon the degree of heterogeneity in relative wealth between and across zones, this is thought to be the best proxy for ability to pay. The range in the value of the zone coefficient indicates the potential level of cross subsidy between users. However, even if the proxy targets poorer households relatively effectively, comparing the change in this indicator with the relative income of the richest quintile and the poorest quintile of the population (see Figure 4.4) shows that the zone coefficient does not accurately reflect the existing levels of economic inequality. The share of income of the highest quintile is 640 per cent of that of the lowest quintile, much greater than the zone coefficient's range.

Moreover, this zone coefficient is only applied in certain parts of the rates system. The zone coefficient is used in Córdoba to determine minimum charges and to assign the adjustment coefficient (AC), reflecting in particular the minimum value of Z (0.8). However, in the Aguas Argentinas concession, the Z factor will only be used temporarily, during the first five years of the concession, to establish free consumption levels for connections with metered service.

Metered Rates

Metering is relatively rare in both cities. Figure 4.5 shows the relative numbers for metered and unmetered connections (residential and non-residential).

The metered consumption system in both concessions is based on a three-pronged tariff, which includes a fixed charge that covers a certain level of free consumption, and a variable charge per cubic metre consumed above the free allowance. In both cases, the fixed charge is associated with the basic rate corresponding to the unmetered system, with a certain distinction between types of users for Córdoba. There are also some differences between the systems where free consumption and variable charges are concerned. As in the case of unmetered consumption, these minor differences can affect the efficiency and equity features of the rates system.

Figure 4.4: First and Fifth Quintiles' Share of Total Income

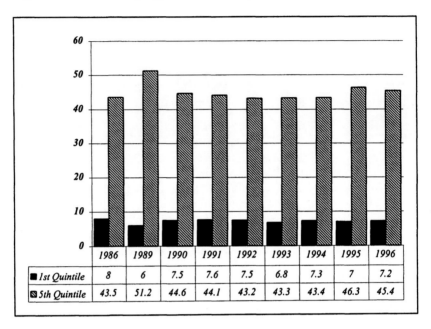

	1986	1989	1990	1991	1992	1993	1994	1995	1996
■ 1st Quintile	8	6	7.5	7.6	7.5	6.8	7.3	7	7.2
▨ 5th Quintile	43.5	51.2	44.6	44.1	43.2	43.3	43.4	46.3	45.4

Figure 4.5: Relative Numbers for Metered and Unmetered Connections

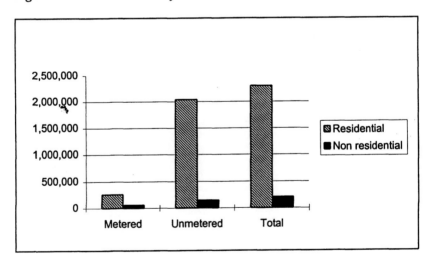

Fixed charge

In both concessions, the fixed charge is expressed according to the minimum charge for unmetered consumption (MIN), and the properties of equity and efficiency as discussed above are maintained.

Table 4.15: Fixed Charges

Buenos Aires	Córdoba
0.5 * MIN	CF(CAT)
	CAT=R -> MIN
	CAT=NR -> 0
	-> 0.7 * MIN

In the Córdoba concession, the fixed charge is only applied for residential users; commercial and industrial users pay a variable charge (see next section). Metered non-residential consumers pay a minimum charge, which is approximately 70 per cent of the unmetered service minimum. The fixed charge for residential customers in Buenos Aires is approximately 50 per cent of the unmetered consumption charge, while in Córdoba it is 100 per cent. It is also offset by different variable charges and free consumption levels.

Variable charge

The variable charges of the two concessions are differentiated - the Córdoba concession makes the distinction between user types, while Aguas Argentinas applies one charge per cubic metre to all types of user. Charges are expressed relative to the AC value.

Table 4.16: Variable Charges (per m³)

Buenos Aires	Córdoba
VC(AC) = $0.33*AC	VC(CAT,AC)
	CAT(R) = $0.28*AC
	CAT(NR) =
	$0.39*AC for regular use
	$0.50*AC for process input
	$1.02*AC for direct input

The distinction in the Córdoba concession is not only by type of consumer (residential and non-residential), but also by use of water. This distinction is based on the observation that the amount of water used in the production process increases from regular use (cleaning and consumption), to an input in a production process (cooling), to being an integral part of the products sold (soda plants). The variable charge applied to residential users is lower in Córdoba than in Buenos Aires, offsetting the relatively higher fixed charge (100 per cent compared to 50 per cent) of the unmetered user rate.

Free consumption

For metered households, perhaps the single most important means of making service payments more equitable is through specified free consumption levels. In both cases this is limited exclusively to residential customers.

Buenos Aires has an interim system effective until the fifth year of the concession which is classified by zone (free consumption increases the lower the economic level of the zone). When this transitional phase ends, a uniform free consumption level of 30 m^3 per connection will be applied.

In Córdoba, free consumption is determined using a decreasing ratio, according to the covered surface of the housing unit.[10] Under this system, the fixed charge depends on the covered area (as is the case with unmetered service) which in turn determines the free consumption volume included in the fixed charge.

Table 4.17: Free Consumption Levels

Buenos Aires	Córdoba
CL(CAT)	CL(CAT)
CAT=R -> 30 m^3	CAT=R ->
CAT=NR -> 0 m^3	CL(SC) = [15 + α* (CS-50m^2)] m^3
R. interim CAT=R (year 1-5)	
CL(Z)	CAT=NR -> 0m^3
Z $\leq$ 1.3 -> CL = 40 m^3	
1.3<Z<1.6 -> CL = 36 m^3	
Z $\geq$ 1.8 -> CL = 32 m^3	

Notes

* α is in the range 0.3-0.1 as SC increases

Conclusion

Given the preponderance of unmetered households in the concession area (and the absence of any real incentives to increase metering), it is exceedingly difficult to devise a rates scheme which is both efficient and equitable. A variety of proxies have been applied. Whether or not these actually reflect relative consumption levels and relative ability to pay cannot be ascertained, but given that the basic rate structure has not been changed for a considerable time it seems unlikely that the proxies are good reflections of actual conditions. Even for metered households, where information requirements are less important, the rate structure seems needlessly complex and flawed. For instance, using the zonal coefficient in metered consumption does not seem to be efficient.

Clearly, a reformulation of fixed charges (eliminating the consumption proxy components), free charges (reflecting the minimum consumption levels the authorities wish to promote for reasons of hygiene and health), and variable charges (so that prices adequately reflect scarcity) will lead to a much more efficient system by pursuing the objectives of equity, efficiency, and sustainability for the concession. It is surprising that the opportunity presented by reform of the sector was not used to revise the tariff structure.

4.5   PRIVATE SECTOR PARTICIPATION AND ENVIRONMENTAL GOALS

Water Allocation and Management

Surface water management

The concept of resource scarcity is not reflected in the structure of rates to be paid by users in the Buenos Aires concession. This is not surprising since the city has a plentiful supply of raw water. However, the rates structure does include the cost of the treatment required to produce potable water, repairs to the water supply network, and water losses from the system.

The plentiful supply of raw water also means that the manner of allocation of water to the concessionaire should not affect water availability for other users (agriculture or industry) or other functions, such as recreation and development of aquatic life. Moreover, industrial development, which is probably the principal alternative consumptive use of potable water, has not historically based its growth on supply from surface sources, as groundwater sources are most commonly used for these purposes.

The abundance of water has led to wasteful water use habits. Indeed, Buenos Aires is by far the largest per capita water consumer in all of Latin America and the Caribbean, and is among the highest water consumers in the world (see Table 4.18). This is even true of metered households. Since water rates are low ($0.27/m^3) and the basic free consumption level is fairly high, the presence of meters *per se* is not effective in reducing consumption levels. While the lack of incentives for conservation does not affect water availability it does mean that Buenos Aires generates high levels of liquid effluents.

Table 4.18: Water Consumption in Major Latin American Cities

City	Water consumption (litres/person/day)	Liquid waste generated (litres/person/day)
Buenos Aires, Argentina	630	96
Mexico City, Mexico	360-527	54
Lima, Peru	359	16
Bogota, Colombia	304	10
Santiago de Chile, Chile	300-555	14
Montevideo, Uruguay	289	6
Sao Paolo, Brazil	270-293	22

Source: Anton (1993) p.156

Groundwater management

In general, groundwater exploitation has not been regulated in Argentina. In Buenos Aires, where use of groundwater is widespread, abstraction rates are not limited in accordance with recharge capacities. Even so, due to the availability of surface waters, groundwater is only a small proportion of the input for the centralised public system. The problems associated with the exploitation of aquifers appears to result from abstraction for industrial users rather than the concession holder's activities.

In principle, one catchment authority should be responsible for administration and regulation of consumptive uses for both surface and groundwater. While this is theoretically true for the Matanza-Riachuelo Catchment Area Committee, the situation for the Rio de la Plata is different. Water use is mainly regulated by the SRNyDS, with the support or participation of other agencies. Owing to the availability of the resource, the regulatory role focuses on ensuring the quality of the resource in connection with its potential uses for conservation and recreation, rather than monitoring levels of water collection and consumption by the concession holder.

Integration of Services

In general, while potential users consider access to potable water to be a priority, sewer connections are not perceived in the same way, since the balance of private and public effects are quite different. While the adverse effects of a lack of access to running water[11] lead mainly (but not exclusively) to private inconvenience and costs, the lack of a connection to the sewerage system is viewed as a public environmental externality that does not affect the homeowner or his/her family to the same degree. Accordingly, people are generally much less likely to request and pay for a connection to the sewerage system. Exceptions can be found in areas where the surface layers are high and constant maintenance of wells and septic tanks is required, leading to the high cost of contracting tank-emptying vehicles.

As in other cities, the advent of potable water in Buenos Aires has normally preceded that of sewerage systems. As a result, while solving one urgent social and health issue, it may exacerbate another equally important one, i.e., the contamination of aquifers resulting from the proliferation of household septic tanks and cesspools. This has an adverse impact on the populations not serviced by the water system but which share the same aquifer.

Another negative repercussion of the non-contemporaneous arrival of these services is that the new volumes of water are transported to areas that do not have adequate drainage. Further, the BAMA includes areas where the surface layers are very high. Thus, providing water services to a neighbourhood where water had hitherto been pumped from the first layer can elevate this layer, leading to significant health problems from saturation and rising sump levels.

This aspect has not changed substantially with privatisation. While potable water expansion for the first quintile reached a total of 1,585,000 persons (26 per cent growth), sewerage reached 810,000 people, i.e., a growth level of some 10 per cent less than water coverage. This can be largely attributed to three factors:

1. Cost of sewer connections: Prior to the renegotiation, a sewer connection cost more than a water connection ($1,500/connection as opposed to $800/connection, respectively);
2. Greater social demand for potable water;
3. The goals established in the Concession Agreement.

Regulating Potable Water Quality

With the prevailing environmental and meteorological conditions in Buenos Aires, conventional treatment technologies are sufficient to meet basic environmental objectives. For instance, the high sediment content in the river water promotes the absorption of substances such as metals. However, when the service was operated by the OSN, there were problems with key water quality variables such as turbidity, bacteriology, and free chlorine. This has been reversed with privatisation. International WHO standards were achieved in 1996 and principal parameters conform to the standards required by the Argentine Ministry of Health.

The concession holder is contractually required to monitor water quality at the level of the distribution system. Since 1993 Aguas Argentinas has applied continuous control to distributed water, from the production plant to the individual home. The relevant sampling and analysis are conducted by the central laboratory of Aguas Argentinas, one of the largest and best-equipped control laboratories in Latin America. The central laboratory is also responsible for general monitoring of the quality of surface and groundwater used. According to WHO recommendations, 70 chemical and bacteriological parameters in potable water are examined. ETOSS concurrently performs ongoing potable water quality checks, and the analyses are performed by the province of Buenos Aires' General Administration for Health Works (AGOSBA) and the University of Buenos Aires School of Engineering.

Regulation of Wastewater Treatment and Quality of Effluents

The Sewerage Plan formulated at the beginning of the agreement did not allow any scope for a substantial improvement in water quality within the concession area (the southern coastal fringe of the Rio de la Plata and the middle and lower catchment areas of the Matanza-Riachuelo and Reconquista Rivers), as the specific needs of different receiving waters were not explicitly considered. This issue has largely been ignored by the State even though National Law 24.051 on dangerous wastes stipulated that SRNyDS should classify receiving waters according to present and future uses.

In addition to modifying the original tariff scheme, the recent renegotiation of the Concession Agreement approved by Decree 149/97 brought major changes in terms of sanitation objectives. In particular, it led to a harmonisation of the Environmental Management Plan for the Matanza-Riachuelo Catchment Area (under the responsibility of SRNyDS) and the Water and Sewerage Plans of the Buenos Aires concession, including "any other issue that contributes to meeting the requirements of the regulatory framework and the needs of public".

Decree 1.167/97 approved the concession holder's proposals for an area sanitation plan, which were designed to reverse the progressive deterioration of the receiving waters in the area. There was a need for rapid classification of receiving waters in accordance with the National Law on Dangerous Wastes. The concession holder defined a new sanitation strategy aimed at ensuring environmental recovery and sustainability of priority uses for the waterways involved. The main differences between the new plan and the original plan are presented in Box 4.1 below.

Both national legislation and regional protocols ("Quality Standards for the Rio de la Plata") have established guidelines for the quality of the receiving water bodies for various uses (consumption, agricultural use, industrial use, protection of aquatic life, and recreational use with or without direct contact). The guidelines do not include a definition of uses for the various receiving waters or sections of such waters. However, SRNyDS recently established uses for the Matanza-Riachuelo catchment area and the coastal fringe of the Rio de la Plata, and the definition of the uses of the Reconquista River are pending. This process was a unilateral initiative on the part of the government, which involved no public consultation. As a result, the uses of receiving water bodies were established without input from users.

Box 4.1: Differences between the original "Sewerage Master Plan" and the revised "Comprehensive Sanitation Plan"

1. **Interception and transportation of contaminated runoff and rainfall.** It is proposed to intercept 17 direct outlets into the Riachuelo River and Rio de la Plata and to pipe this water into a purification plant. This was not considered in the original plan.

2. **Adaptation of treatment plants to the environmental characteristics of receiving waters.** Given that each receiving body (Matanza-Riachuelo, Reconquista, and de la Plata Rivers) has its own specific morphological and hydrodynamic characteristics giving it different assimilative capacity, the sanitation plan proposes to adapt the number of plants, the volume of discharge and types of treatment to the acceptability of the receiving environment in question. This issue was not considered in the original plan.

3. **Flexibility of the collection system and plants.** The collection system planned for Berazategui within the original Plan is now divided into two subsystems, generating operational flexibility and greater security in the overall collection system, which is less dependent on the existing *Cloacas Máximas* transport system.

4. **Adaptation of treatment and disposal of biosolids by plants.** In light of the new implementation standards, the CSP considered complete sludge treatment in three of the plants. Two designated sanitary landfill disposal sites are provided.

5. **Speeding up expansions.** Bringing forward expansion of service for 240,000 people (from the third to the second five-year plan) in the Matanza-Riachuelo catchment area, according to the environmental management plan for that area.

6. **General system monitoring.** The CSP proposes implementing an overall system monitoring network: to ensure proper hydraulic functioning of the collection system and to monitor the quantitative and qualitative changes in effluents intercepted; to monitor adequate operation of plants and the quality of their effluents; to monitor the effects of sludge discharge sites on their environment; to verify the appropriacy of the guidelines proposed in priority use areas; and to monitor the general quality of the Rio de la Plata in the area of the concession.

Generally speaking, this means that not only must users accept the specified uses, but they must also directly or indirectly bear the cost associated with treatment of effluents.[12] These unilateral designations of quality standards are often based on very ambitious objectives copied from standards developed for other countries with different environmental and socio-economic conditions. It is important for the public sector to review the procedure that is being used to determine environmental standards. In particular, participatory approaches that allow for discussions among the sectors involved should be used. Of course, some of the reasons for doing so are economic ones. Users will not necessarily support decisions involving the absorption of higher costs for environmental improvement unless they have an input. Opinion surveys in Argentina show that, while a large proportion of the population would like an improved environment, not everyone is prepared to pay for it.[13]

Table 4.19: Monitoring Results for Industrial Discharges to Sewers

Industry	Sample	Number in violation of discharge standards	Numbers exceeding discharge standards for specific pollutants				
			TSS[1]	BOD[2]	COD[3]	Deterg.	Chrom.[4]
Meat packing	155	144 (93%)	117	98	126	22	0
Other foods	113	93 (82%)	59	50	71	24	0
Textiles	236	210 (89%)	111	80	75	142	7
Tanneries	67	62 (93%)	36	29	45	3	54
Soaps	16	14 (88%)	14	5	10	4	0
Other chemicals	44	36 (82%)	14	17	21	9	7
Metals/ machinery	91	61 (67%)	11	2	13	3	38
Others	41	29 (71%)	1	16	4	8	-

Notes
1 TSS (Total Suspended Subst.)
2 BOD (Biological Oxygen Demand)
3 COD (Chemical Oxygen Demand)
4 Chrom. (Chromium)

Source: World Bank (1995) from data provided for Aguas Argentinas S.A. for the period 1993-94.

Industrial effluents discharged outside of the sewerage system were not part of the concession holder's obligations under the original contract.[14] However, Aguas Argentinas is required to monitor the quality of waste discharged by industry and other establishments into the sewerage system and to notify SRNyDS of violations of the established standards. The concession holder began this monitoring in 1993 and the results clearly show large-scale non-compliance. Table 4.19 provides a summary of the data from 1,140 inspections, in which only 167 establishments met standards for all parameters covered by the regulations.

The risk of illegal discharges into the sewerage system is reinforced by the existence of numerous micro-, small-, and medium-scale industries throughout the concession area. These firms have limited access to capital, technology, information, and markets. Accordingly, their capacity for appropriate environmental management is quite limited. The principal constraints on reducing environmental impacts from smaller firms can be summarised as follows:

- Ineffective regulatory enforcement encourages micro- and small-scale industries within the serviced area to discharge their effluents into the system;

- While the concession holder can verify and report violations, it has little authority to police or levy penalties;

- Sewers and storm drains transport industrial liquids that are discharged untreated (except in the case of the South Western plant at present, and Northern plant in the near future) into receiving waters, contributing to the pollution of the Rio de la Plata and the Berazategui outlet channel;

- Drainage of industrial wastewaters through conduits increases the risk of treatment plants going out of service because of damage from certain substances at various stages of the process.

Privatisation of the sector will only have a relative impact on environmental conditions in the area since other sectors (such as government and industry) contribute to environmental degradation. Under the present circumstances, the concession holder does not have sufficient authority or incentive to improve pollution control. Can the concession holder's present role in industrial control be changed, allowing it to take direct action against industry for dumping into the sewerage system? For instance, consideration could be given to granting the concession holder authority to collect fines (as other privatised services do), not only on late payment, but also for violations of environmental regulations (prohibited discharge, etc.).

Conclusion

Greater water and sewerage system coverage, gradual elimination of sumps and septic tanks, and adequate treatment of liquid sewerage could produce the positive externalities that are essential for improving the living conditions of the population in the concession areas. These externalities can be summarised as follows:

- Improved quality of receiving waters;

- Improved quality of groundwater;

- More rational use of aquifers;

- Upgraded effluents and wastes for other uses (e.g., biosolids in agriculture);

- Improved quality of human life and the environment in general.

In Argentina, household effluents have traditionally not been treated, and only a few major cities, such as Córdoba and Mendoza, have functioning treatment plants. The implementation of the CSP will be a turning point for environmental quality in the BAMA. In addition, increased service coverage will reduce unregulated groundwater abstraction, unregulated wastewater discharges into surface waters, and unregulated seepage from septic tanks and latrines into groundwaters.

Despite these potential benefits, the CSP has not been subject to an environmental impact assessment (or a strategic environmental assessment, which would be more appropriate). The effects of pollution on the receiving water bodies should be incorporated into such assessments. In the past few years, the concession holder has made tremendous qualitative contributions compared to the former OSN in terms of conducting technical studies. However, it is unfortunate that a comprehensive overview of potential impacts of the sewerage plans and CSP is still not available.

One of the most important benefits of providing adequate sanitation and potable water is the impact on waterborne diseases (diarrhoea, gastroenteritis, cholera, etc.). In this connection, disadvantaged households are most vulnerable to the impacts of pollution (contaminated aquifers and surface waterways, contaminated food, etc.), and the quality of life for this sector is most rapidly improved through the availability of such services (see Cairncross and Kinnear 1992 for a discussion of the health benefits of improved access to adequate water and sanitation facilities). Therefore, the next section will review the particular implications of the concession for poorer households.

4.6   PRIVATE SECTOR PARTICIPATION AND SOCIAL OBJECTIVES

Access to Services for Disadvantaged Households

Access to potable water and sanitation is an "essential" good, and adequate access is a pre-condition for people to participate fully in society. While Argentina has the highest per capita GDP in Latin America, it has one of the region's lowest coverage rates for these services. This is due in large part to the lack of investment by the former OSN, which prevented network expansion.   Notably, there was very low coverage in economically disadvantaged neighbourhoods (and almost no coverage in the very poorest neighbourhoods) at the beginning of the concession (1993).

Low income population in the concession area

It has been estimated that 60 per cent of the 1,800,000 people who do not have access to potable water are below the poverty line.   On the basis of factors such as urban planning, land tenure, social structure (neighbourhood organisation and social stability) and the economic situation of the population (income and unemployment), three general "types" of disadvantaged neighbourhoods or settlements can be identified:

1.   *Villas miserias* (squatter settlements). These include informal urban settlements without planned streets or footpaths and where houses are often made of makeshift materials. Residents generally do not own the land. Average total household income does not exceed $350 per month. The basic needs of the majority of the population are unmet (*necesidades básicas insatisfechas - NBIs*). Such settlements are frequently located on degraded land prone to flooding, close to contaminated waterways, with little access to basic services and poor accessibility by road.

2.   *Barrios humildes* (disadvantaged formal neighbourhoods). These neighbourhoods range from formal settlements with adequate street planning to those in which as much as one-quarter of the population are *NBIs*. Dwellings are often owner-occupied and located in less environmentally degraded areas, with housing largely built with permanent materials. Although they have streets and footpaths they are not always paved. Average household income varies from $350 to $650 per month. There are persistent problems with provision of services and basic infrastructure.

3.   *Complejos habitacionales* (housing complexes). These include public housing blocks, which began to be built in the 1960s to house the

underprivileged. Serious maintenance problems in such housing often makes them uninhabitable. The socio-economic situations of the inhabitants vary from those in the *NBI* category to owner-occupiers.

The degree of access for each of these groups in the concession area is presented in Table 4.20.

Table 4.20: Service Access for Underprivileged Population in the Concession Area

Total underprivileged population	Within service area	Outside service area	Total
Population in *villas miserias*	**250,000**	**50,000**	**300,000**
Formalised	100,000	0	100,000
To be formalised	150,000	50,000	200,000
Population in *barrios humildes*	**410,000**	**1,100,000**	**1,510,00**
Formalised	320,000	0	320,000
To be formalised	90,000	1,100,000	1,190,000
Population in *complejos habitacionales*	**110,000**	**90,000**	**200,000**
Formalised	110,000	90,000	200,000
To be formalised	0	0	0
Total	**770,000**	**1,240,000**	**2,010,000**
Formalised	530,000	90,000	620,000
To be formalised	240,000	1,150,000	1,390,000

Source: Aguas Argentinas S.A./IIED-AL (1998)

Connection charges, network expansion and poorer neighbourhoods

Given this bias in access, one of the key factors in determining the distributional effects of the rates system relates to the financing of system expansion. The original financing system established an infrastructure charge, with each new user paying the full cost of a new connection.[15] This, together with the considerable expansion needs and the socio-economic characteristics of the population in the expansion areas, led to a series of problems in Buenos Aires, and hence the need to modify the rates system.

In many neighbourhoods, the infrastructure charge represented a large proportion of average annual income. To analyse the impact of costs of services on disadvantaged households, the infrastructure charge should be distinguished from the regular service payment. While the service payment[16] accounts for 1 to 2 per cent of a poor resident's monthly income, the initial system connection charge was $600 per family for water and $1,000 for sewerage - unaffordable for a low-income family.

In light of the inequities in access to formal systems and the high initial costs of connection, many poorer families decided not to connect to services. In other areas, there was a high rate of non-collection for new connections, and this impacted on the economic and financial sustainability of the concession. Clearly, the Buenos Aires concession was not designed for the bulk of the expansion to be in areas in which there is a high proportion of poor residents. The inability and unwillingness of households to pay the charge has led to considerable financial shortfalls for the concessionaire.

This situation reached a crisis in mid-1997, leading to a reformulation of the expansion rules (Decree 1167/97) and the elimination of the infrastructure charge. Under the new system, charges for new connections, which had been paid by each user, were replaced by a universal service charge to be levied on all network users. Accordingly, the concession holder set out to renegotiate the contract in order for the concession to be financially solvent. This renegotiated arrangement incorporated the rate increase, as well as the development of services not initially provided for in the Comprehensive Sanitation Plan for the Buenos Aires Metropolitan Area.

The charge had to be sufficient to ensure that the company's financial position was not altered. Therefore, the net present value of projected revenue from the infrastructure charge had to equal the net present value of revenue generated by the new charge.[17] To ensure parity with existing users who had already paid some or all of the infrastructure charge, these payments also had to be reflected in determining the universal service charge. Based on the new rates agreement, the costs were $120 for a potable water connection and $120 for the sewerage system (to be paid in five years at a rate of $4 every two months) which also included the service. This scheme was not generally favourably received, and, once again, the fact that the concession authority, the SRNyDS, did not consider other alternatives such as direct subsidies to the underprivileged population, is being criticised.[18]

While the new charge is progressive, insofar as households with new connections (which tend to be disproportionately poor) are cross-subsidised by households with existing connections (which tend to be disproportionately rich), there are other distributional issues. The universal charge is a fixed charge per user with no distinction of any kind, which has meant a reduction in cross-subsidies between commercial and residential users. On the other hand, it more than proportionally increased costs for users with minimum invoices (representing more than 35 per cent of total users). While this undermines some of the progressivity of the rate structure, it is important to bear in mind the relatively low level of existing minimum charges ($4 every two months).

Although connection costs have been substantially reduced, it remains to be seen how the additional costs that a disadvantaged family will have to pay to adapt the internal infrastructure of their home will be financed. These costs are approximately $50 for water and $400 for sewer connections. The question arises as to whether they will be able to bear these costs, or if other sources of financing should be considered, such as the concession holder, the government (directly or through multilateral credit), or NGOs through international assistance.

While this change in the system was approved by the authorities and accepted by the concessionaire, its implementation was subject to a legal dispute. A judge of first instance issued an order to suspend the charge based on the argument that it is a tax and as such the executive power does not have authority to establish it, as only parliament has this exclusive right. However, this has since been resolved.

Informal collection and the centralised system

The privatisation of Buenos Aires' services revealed the magnitude of the investment shortfall in WSS infrastructure. Moreover, with the establishment of explicit goals, including universal provision of services, it quickly became evident that service conditions varied for different social sectors. Most disadvantaged neighbourhoods include users who are connected illegally to existing conduits, within or outside of the area serviced with potable water. This generally results in a break or deterioration in the infrastructure, leading to water loss from the centralised system.

This system of illegal collection of potable water from the centralised system, is sometimes organised as a parallel system, with households paying a water "authority" for access. These parallel illegal networks often present serious problems in terms of water quality, irregular supply, and low pressure. However, with the execution of the *Plan de Acción de Barrios Carenciados* ("Action Plan for Disadvantaged Neighbourhoods") for the BAMA,[19] almost all of the illegally connected consumers (as many as 600,000 at the beginning of the Concession Agreement) were incorporated into the centralised system.

There are other systems that are not illegal, but which arose out of independent projects developed before services were privatised, using national or provincial funds (for example, ProAgua and ProviAgua plans), or third party works systems (OPCTs) paid for by the residents themselves. These parallel systems are generally operated by neighbourhood co-operatives, grassroots organisations, or local development organisations. They often face the same water quality and supply problems as the illegal systems. Many of these have also been incorporated into the centralised system.

Opinion surveys[20] among residents who have switched from an informal provider to service provided by the concession holder show a high level of satisfaction with the new system. Field studies conducted by IIED-AL in a number of townships in the concession area in 1997[21] found that lower-income consumers registered higher satisfaction levels with the new service.

This is not surprising given the qualitative leap that this change represented for the disadvantaged sectors.

The Role of Users and Public Education

During the past two decades, the top-down approach in water and sanitation projects targeting disadvantaged urban populations has been found to be quite inefficient. As affirmed by Wright (1997), "... the supply-driven approach of the traditional sanitation agenda implies that users are not consulted about the services that they want and for which they are willing to pay. This means that potential consumers simply ignore the resulting systems ..."

The role of users has changed in a number of ways since privatisation. It could be said that the concession is more customer-oriented than OSN had been. The enterprise is required to meet a series of service quality standards and implement customer service improvement. In this respect, new service and payment points have been created and special lines have been established for technical and commercial complaints, as well as other facilities.

The time required to address service complaints has been substantially reduced. Under OSN, complaints were required to be dealt with within five to 15 days, but they are now resolved within a maximum of three days. Complaints or queries not receiving proper follow-up through the company may be referred to ETOSS. In practice, and particularly with reference to rates increases, complaints are filed through civil consumer organisations and the ombudsman (*Defensor del Pueblo de la Nación*). These groups in turn file complaints with the regulatory agencies, the enterprise, or the justice system.

However, as far as the regulatory structure is concerned, the role of users is focused on control of services, which is entirely complaint-driven, with no other type of active user participation in design, planning, service development, infrastructure maintenance, or system operation. There is no participation or formal direct representation of users in agencies such as ETOSS. In principle, the role of users could take on a quite different dimension where disadvantaged users are concerned.

The active participation of users in the design, implementation and management of WSS projects in many underprivileged neighbourhoods, especially the poorest ones, is constrained by the overall socio-economic and cultural environment in which they live. Any policy to provide services should be accompanied by other strategies and actions, for example, in the area of education. It is not clear how such responsibilities should be distributed. Should the concession holder bear responsibility by incorporating an education component into the provision of services or should the government retain control of this function?[22]

Once again, the best approach might be to develop joint strategies between the concession holder and the concession authority, incorporating active participation of grassroots organisations and locally active NGOs. Indeed, this type of approach is consistent with Agenda 21, according to which schemes must be presented to enhance participation and inter-sectoral integration in order to promote sustainable development of societies.

Criteria for Expansion Areas

Expansion of services has traditionally been dominated by technical and economic criteria. In other words, even during the former OSN administration, areas of expansion were basically chosen according to the proximity to existing major conduits or collectors, and the existence of conditions that would permit a reasonable connection to the centralised system (technical feasibility), and depending on the level of investment required and the local population's capacity to pay for services (economic feasibility).

Box 4.2: Optimisation of Expansion Plans in Greater Buenos Aires

Based on the framework agreement signed by Aguas Argentinas and IIED-AL, a study was conducted to identify relatively homogeneous areas of socio-environmental demand for services in 19 townships in the expansion area of the concession. This study involved the establishment of four indictors:

- Social stratification index (IES)
- Potable water access deficit (DAP)
- Sanitation access deficit (DAS)
- Groundwater contamination (CAS)

The objective of the study was to help identify feasible alternative solutions from technical, economic, and institutional standpoints, to address the situation in the most critical areas. The study included an analysis of the characteristics of the communities involved, their levels of social organisation, and the existence of community leaders and grassroots organisations present in the neighbourhoods. Table 4.21 presents new feasibility criteria, based on this new expansion programming method, which the enterprise began to use to implement the Action Plan for Disadvantaged Neighbourhoods.

As a result of an agreement signed between Aguas Argentinas and IIED-AL (see Box 4.2), efforts were made to incorporate social and environmental variables in order to reorient the expansion programme. This new scheme reflects the existence of critical areas (with greater social and environmental demand) near and within the serviced area. It also makes it possible to plan and design feasible solutions for those critical areas whose remoteness from the serviced area made them candidates for coverage by alternative or non-conventional systems and sources (such as pumping from deep, high-quality layers, pumping and treatment, pipe transport through underground waterways).

With this new scheme, traditional methods for the expansion of the system can be combined with alternative interventions or systems described below. Arrangements such as compensation agreements with townships or third party works would make it possible to expedite service coverage in those areas where the concession would not provide services for many years. The success of this approach highlights the need to change the present conditions of the Concession Agreement. This change would theoretically require the degree of contractual rigidity to be reconsidered in light of the benefits of greater flexibility and adaptability. This would enable a continuous revision of goals and objectives, essential to adapt coverage goals to requirements in connection with rapidly changing social and institutional dynamics.

Table 4.21: Technical, Social, and Institutional Feasibility of Network Expansion in Disadvantaged Neighbourhoods

	Variable	**Indicator**
Technical Feasibility	*Location*	Proximity to potable water source
	Land tenure	Eradication -Non-eradication Expropriation in progress Government or private land (legality)
	Physical and environmental conditions	Flooding, Health risk, Soil types
	Urban planning	Regular - Irregular
Social Feasibility	*Social structure*	Level of social organisation Level of social stability (security/existence of references/alcoholism and drug use) Social capital (confidence, solidarity, reciprocity) Education level
	Economic situation	Income level, Unemployment level
Institutional Feasibility	*Political and institutional structure*	Level of township participation Level of co-ordination among public agencies Co-ordination between political figures (neighbourhood leaders, town councillors) Relations between government sectors (National/ Provincial/Township)

Specific Solutions and Strategies for Underprivileged Neighbourhoods

Despite the constraints of the regulatory structure, Aguas Argentinas has been willing to incorporate technological innovations as well as participation of local communities, neighbourhood organisations, NGOs, and local governments into water and sanitation system projects. This has enabled them to:

- Overcome physical and environmental restrictions in poor settlements located on landfills, floodplains, or highly contaminated areas not suitable for the development of traditional infrastructure;

- Anticipate the provision of services when critical neighbourhoods are very far from the serviced area, particularly in areas where services are not going to be provided for more than 15 or 20 years;

- Reduce investment costs to facilitate the provision of services for these cases;

- Promote an attitude of ownership among the target communities to foster conditions of security during the execution of the work and subsequent maintenance stages, as well as adequate conditions for payment for the service, and effective integration of disadvantaged households into the formal system;

- Promote an environment of co-operation and effective relations between the service operator, the communities involved and local governments to ensure the long-term sustainability of the systems implemented.

On the basis of the Buenos Aires Concession's experience, it is possible to identify different strategies which imply not only the use of appropriate technological solutions[23] but also alternative institutional approaches. On the one hand, innovative technological solutions are very important to allow the provision of services in complex environments, for example where lands are unstable, highly polluted or subject to drainage problems. On the other hand, institutional solutions are based upon different models of participation of disadvantaged communities and local government.

Various experiences in the concession area can be characterised according to the technologies used, the participants, type of agreement, size of the population involved and type of internal working organisation. Box 4.3 provides the principal institutional strategies used in the Action Plan for Disadvantaged Neighbourhoods in Buenos Aires.

Box 4.3: Institutional Strategies for Disadvantaged Neighbourhoods

A. Water service by consensus

Participants - Type of contract	Tripartite contracts - Concessionaire/ Municipality / Neighbourhood
Population range	500-2,500 persons
Internal organisation	Exchange of labour for connection
Example	Barrio San Jorge (San Fernando) Aguas Argentinas

B. NGO Intervention

Participants - type of Contract	Quadripartite contracts: Concessionaire/ Township / Neighbourhood / NGO
Population range	5,000-15,000 persons
Internal organisation	Participation of an NGO to co-ordinate relations between the participants
Example	Villa Jardín (Lanús) / NGO: Fundación Riachuelo

C. Compensation Agreement

Actors - Type of contract	Bipartite agreements: Concession holder/ Municipality
Population range	Up to 6,000 persons
Organisation	Compensation for connection charge through municipal fee and tax exemptions
Example	Barrio Santa María (Municipal de Lanús)

D. Job Creation Unit

Actors - Type of contract	Tripartite agreement: Concession holder/Province of Buenos Aires/Municipality
Population range	Up to 50,000 persons
Organisation	The provincial government pays the cost of materials and advances the cost of labour.
Example	Villa Albertina (Lomas de Zamora) - 50,000 persons

These strategies have had varying degrees of success. The intervention of NGOs has proved to be successful, and has achieved the desired goals. Others, such as job creation units, have had mixed experiences. Even though they have allowed for a considerable increase in service coverage in poorer areas, their success is contingent upon political and institutional factors and they do not offer a stable future. The different strategies can therefore evolve into other approaches, provided that acquired experience and lessons learned are used constructively, and technologies and management mechanisms are optimised to ensure the long-term sustainability of the systems to be implemented.

In the interventions outlined in Box 4.3, NGOs enable co-ordination among the three parties and can play an important role as facilitators or catalysts in the processes.[24] The objective was to change the working relationship from a top-down to a participatory and strategic approach. Such an approach is a way to take advantage of the institutions' different capacities and exploit complementarities. In this scheme, each institution provides that part of the service for which it has a comparative advantage.

Each sector has a different role. The company is responsible for the project design and supervision during the implementation phase. The local institution (municipality), which is the public authority, provides construction materials. The local community is responsible for providing participation from its members. In this scheme, the NGO can provide social training for the company, co-ordination and technical assistance for the public institutions, and internal organisation and capacity strengthening for local communities and their leaders.

The transparency and credibility of the NGO is vital, particularly for the local communities. Box 4.4 provides an overview of the results of the implementation of the Action Plan for Disadvantaged Neighbourhoods in Buenos Aires.

Box 4.4: Aguas Argentinas - Action Plan for 1997

The following gives the number of people involved in the first stage of Buenos Aires potable water expansion (1997-2000), and the low income inhabitants included in the Action Plan for 1997.

Number of people included in expansion plans for 1997-2000	1,130,000
Low income settlements (population) involved	751,300
Low income inhabitants in the Action Plan for 1997-98	190,600
Formalised in 1997	110,200
To be formalised in 1998	80,400
To be formalised with future expansion	558,100

Through this project two significant objectives can be realised: co-operation between different actors and sectors; and improvements in the livelihoods of the most vulnerable members of society.

Conclusion

The implementation of appropriate solutions for the specific local conditions and the social organisation of the communities (for example, Barrio San Jorge, San Fernando) demonstrates that even the poorest populations can participate in programmes to provide infrastructure and services, as long as they are within their financial means. It should therefore be possible to design a system differentiated by technical, economic, and institutional factors that will expedite the provision of services in cases where, for environmental or social reasons, services are urgently required. This involves partnership amongst the various groups, with each one carrying out functions in accordance with its skills and experience.

4.7   CONCLUSIONS AND POLICY RECOMMENDATIONS

Initial Scenario

During the past decade, public potable water and sanitation service operators have confronted various problems, the most serious being lack of investment capacity and cash flow. In Argentina in 1992 (before privatisation), OSN spent more than 80 per cent of its budget on operations and less than 5 per cent on new investment. This meant that there was very little service expansion (particularly in Buenos Aires) or maintenance.

This situation had four main social and environmental consequences:

1. **Highly polluted surface and groundwater**. An estimated 2.2 million m³/day of untreated household wastewater and 1.9 million m³/day of industrial effluents flowed through the BAMA to the Rio de la Plata, the population's main source of potable water. Sumps used by households without sewer connections (more than 6 million people in Greater Buenos Aires) led to groundwater pollution;

2. **Economically disadvantaged families were least likely to have access to services**. A high percentage of the population without access to safe water (4.5 million) or appropriate sanitation (6.5 million) in Argentina is poor.

3. **Water and service quality problems for connected households**. Insufficient pressure, turbidity, and bacterial contamination were recurrent problems;

4.   **Inefficient use of water resources.** There were high levels of waste and
     water loss (unaccounted-for water).

Government agencies had few of the supervisory authorities that concession
holders now have (concession authorities, regulatory agencies, and audit
agencies), with supervision limited to ad hoc activity by the non-
governmental sector, the press, and, to a lesser extent, multilateral credit
institutions. The absence of clear objectives and explicit contractual goals for
OSN also hindered supervisory activity. Consequently, there was a lack of
awareness and institutional support for adequate allocation, management, and
rational use of water resources, since there was little information on critical
environmental and social issues, particularly with regard to the impact on
public health.

The Impact of Privatisation

The privatisation resulted in significant changes, particularly:

- Increased investment capacity and cash flow, essential for renewed
  expansion in the sector and for rehabilitation and operational
  maintenance of the existing infrastructure;

- Improvement and rehabilitation of potable water and sewerage systems,
  which were on the verge of collapse in many areas;

- Commencement of construction on the Northern Wastewater Treatment
  Plant in the township of San Fernando and planning of other major
  works. (Progress has been disappointing as the concession holder was
  required to carry out unplanned rehabilitation work on the systems);

- After decades without access to safe water or adequate sanitation, service
  is now provided for 110,000 people living in *villas miserias* and other
  disadvantaged neighbourhoods within the serviced area by means of
  various technological and institutional solutions within a framework
  based on the participation of the communities, local governments,
  NGOs, and grassroots organisations;

- Tariffs remain below their pre-privatisation levels. By some measures
  water is the least expensive service in Buenos Aires;

- Improved service quality, particularly that of water for distribution.
  Levels for turbidity, bacteriology, and free chlorine comply fully with
  international standards;

- A significant reduction in the level of water loss (from 45 per cent at the start of the concession to less than 35 per cent in year five);

- A clear qualitative improvement in operator capacity has enabled the concessions to undertake important technical research, including environmental studies on groundwater, environmental evaluations, environmental impact assessments, and social surveys for the purpose of scheduling and optimising the concession holder's activities.

With the elimination of the infrastructure charge in 1997, a process of renegotiating the Concession Agreement was initiated. The concession holder and the authorities agreed on new tariffs and a new sanitation plan for the BAMA, which was more ambitious than that provided in the original agreement. However, this is still awaiting approval.

Pending Issues

There are a number of other problem areas which need to be addressed as a matter of priority:

- Many disadvantaged neighbourhoods located outside of the currently-serviced area will not receive services for ten, 15, and even 20 years;

- The present contractual arrangements hinder the implementation of non-traditional solutions or those designed to expedite the arrival of services where they are needed;

- Water provision and sanitation services are provided separately and sanitation services are still less accessible than water supply;

- A clear strategy for provision of sanitation services in the area has yet to be defined;

- Although the tariff system complies with principles of equity, it provides no incentives for expansion to disadvantaged areas or for efficient water use;

- There are no clear provisions for public health and education issues. The expansion initiative makes no explicit provision for them, and the government has left it to the concession holder to address;

- Privatisation will have a limited impact with regard to the improvement of environmental and living conditions in the area, since other sectors (such as government and industry) are not contributing to the control of pollution;

- While the concession holder may report illegal industrial dumping, it does not have the power or incentives to exercise effective control in this area. Given that government control of industrial contamination is ineffective, no solution to this problem can be envisaged in the short or medium term.

Public Policy Recommendations

The following recommendations are designed to contribute to improving the political and institutional conditions in which water and sanitation services are provided in Argentina. Case studies in other countries of the region, that share many social and cultural factors and that are undergoing similar public and private sector participation processes, will also provide further insight.

- Promotion of the development of technologies and appropriate institutional approaches, where necessary, with the aim of providing universal and comprehensive access to potable water and sanitation services;

- Implementation of flexible regulatory schemes that enable the application of appropriate innovative institutional and technical solutions when traditional approaches prove insufficient or difficult to implement;

- Implementation of environmental management systems (pursuant to ISO Series 14000 standards) to prevent the negative environmental externalities associated with the activities of the enterprises operating the services, thus improving the environmental and social performance and efficiency of privatised services or concessions;

- Development of policy dialogue and co-operation programmes between the government, industry, and the concession holder, to find integrated, consensual solutions for the reduction of water pollution levels. Small and medium-scale industry and the concession holder should be provided with adequate incentives, including fiscal instruments and multilateral financing;

- Implementation of tariff systems and market-based instruments that provide appropriate incentives for the private sector to achieve environmental and social goals;

- Interaction between the government and the concession holder to develop educational programmes and to optimise the impact of the services on public health. The general health conditions of the poor are related to the overall environmental, socio-economic, and cultural context in which they live, and any policy to provide services should include education strategies;

- Development of strategies based on the participation of local disadvantaged communities in water and sanitation projects from their commencement;

- Promotion of approaches based on the integration of public service bodies, government institutions, and disadvantaged communities, with the support and facilitation of NGOs or neighbourhood associations, to promote the long-term sustainability of the water and sanitation systems being implemented, while ensuring appropriate levels of operational efficiency.

The transformation of the initial supply-driven system to a demand-based system, in which adequate political, institutional, and regulatory conditions are created in the best interest of society at large (incentive-driven), should be expanded. Implementation of these recommendations on public-private strategies may prove effective in achieving the goals of social equity, ecological sustainability, and political governability.

ANNEX 1: AGUAS ARGENTINAS RATES SYSTEMS

The rates system incorporates a metered consumption system with scope for a fixed charge to be applied to some types of user. The metered charge system is mandatory for non-residential users and "bulk" water sales, and is optional for residential users. The concession is subject to a price regulation system based on price control with a registered cost mechanism. The concession holder is free to establish lower rate levels and prices, provided that it does not apply reductions on a discriminatory basis. Cost variations exceeding a pre-established level constitute a rates revision.

The concession holder is required to grant, at its own expense, the exemptions provided for religious organisations and foreign representative offices (Organic Law of the OSN, Article 71 (a) and (b)), and for fire-fighters. Any other existing or future exemption, reduction, or subsidy (retired persons, public welfare institutions) will be the responsibility of the concession authority, which will earmark specific allocations in the national budget for that purpose.

In the event of arrears, the concession holder will be entitled to apply a 5 per cent surcharge to the original amount billed during the first month. After the first month, the penalty charge will increase to 10 per cent (though not on an accrued or additive basis with the preceding charge). After the first month, the concession holder may also collect an indemnity in addition to the original amount, equivalent to 1 per cent of the cumulative monthly amount. This system may be amended by the regulatory agency.

In cases of late payments in more than three billing cycles, the concession holder will be authorised to discontinue services, subject to at least two written warnings and a minimum of seven days' advance notice. Following collection, the concession holder must restore service within 48 hours. To protect public health, the concession holder is not permitted to discontinue service to public or private hospitals or clinics. If these institutions fail to comply, the concession holder will refer the matter to the regulatory agency, which will take the necessary steps to obtain payment within a maximum of six months.

If a user's service is erroneously discontinued, the concession holder must restore service within 48 hours. Should the concession holder fail to do so, it will compensate the user with a payment equivalent to 30 per cent of the total invoice during the last complete bi-monthly period for every additional day.

Rates System

The rates system includes three types of buildings: residential, non-residential, and undeveloped or derelict property. Buildings in the first two categories are divided into single-unit and multiple-unit serviced buildings.

The five categories established using this procedure are given in the table below:

	Single Unit	Multiple Units
Residential	R I	R II
Non-residential	NR I	NR II
Derelict property	Derelict property	

Methods for applying water and sanitation service charges include metered or unmetered consumption. Buildings without consumption meters must pay a fixed bi-monthly charge calculated on the basis of the covered surface of the building and 1/10 of the land area. This yields the Basic Bi-monthly Rate (BBR).

Buildings fitted with consumption meters are levied a fixed charge plus a variable charge per cubic metre consumed, as explained below.

An explicit objective of the regulatory framework is for the basic rates system to be based on a metered consumption system (Article 45) and to that end, it stipulates that the installation of meters is mandatory for all non-residential buildings and "bulk" water sales. Users are required to pay installation charges (Article 49 of the Regulatory Framework).

If the concession holder is unable to apply the metered rates system immediately, on a one-off basis, ETOSS will allow the concession holder a period of two years to comply. During this period, the fixed-rate system (BBR) will apply. At the end of the period, the concession holder will incorporate all non-residential users, and in the absence of actual metering, it may only bill the appropriate fixed charge for each category.

Residential customers may have meters fitted on an optional basis. This option may be elected by users or by the enterprise, and the costs are borne by the party who decides. The basic bi-monthly rate is determined according to the general rate (GR), geographic zone (Z), adjustment coefficient (AC), covered surface (CS), type and date of construction (C) and land area (LA). This rate is calculated using the following formula:

$$BBR = AC . Z . GR . (CS . C + ST/10)$$

This basic bi-monthly rate is subject to minimum levels (discussed below).

GR: General Rates (Article 14)

General rates are divided into building categories according to the following table:

Building	Water $/m^2$	Sewerage $/m^2$
Residential	0.0279	0.0279
Non-residential	0.0558	0.0558
Derelict property	0.0279	0.0279

Within the old perimeter of the federal capital, these rates will be 10 per cent higher than the current levels, as storm sewers are combined with sewers in this area.

C: Type and date of construction (Article 15)

The C coefficient has two components (type of building and average date of construction) which are designed to reflect the user's payment capacity.

This coefficient increases with the building construction date and building category according to predetermined values as illustrated in the table below.

C Coefficient by Date and Type of Construction

Const Date	< 1932	1933 1941	1942 1952	1953 1962	1963 1970	1971 1974	1975	1976 1986	1987 1992	1993 2002	2003 2012	2013 2022	> 2022
Lux	1.62	1.68	1.75	1.82	1.90	1.97	2.04	2.35	2.65	2.91	3.21	3.53	3.88
VG	1.47	1.52	1.53	1.65	1.72	1.78	1.85	2.13	2.40	2.64	2.90	3.19	3.51
Good	1.25	1.29	1.34	1.40	1.46	1.51	1.57	1.81	2.04	2.24	2.47	2.72	2.99
G LC	1.07	1.10	1.15	1.20	1.25	1.30	1.34	1.54	1.74	1.91	2.11	2.32	2.55
LC	0.89	0.92	0.96	1.00	1.04	1.08	1.12	1.29	1.45	1.60	1.75	1.93	2.12
VLC	0.64	0.66	0.70	0.72	0.75	0.78	0.81	0.93	1.05	1.16	1.27	1.40	1.54

The construction quality component is designed to reflect differences in the user's capacity to pay for the service. The quotient between different categories remains constant over time, with only minor differences (hundredths). The ratio between extremes is 2.5 to 1. Using the coefficient from the "good low-cost" construction category as the base, we obtain the following implicit subsidies and excess costs in this coefficient:

Luxury	1.51
V Good	1.37
Good	1.17
Good LC	1.00
Low-Cost	0.83
Very LC	0.60

The date of construction component of the variable C shows a particular evolution. After remaining at a constant annual change level of about 0.45 per cent per year for houses built between 1932 and 1975, for houses built between 1975 and 1989 the coefficient rises by approximately 2.3 per cent per year, and then declines to around 1 per cent per year beginning in 1992 (see graphic).[25]

C Coefficient Variation by Year

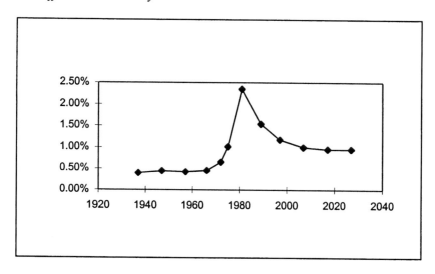

Z: Geographic area (Article 16)

As with construction type, this variable aims primarily to reflect payment capacity. The values of this coefficient vary between 0.8 to 3.5. Present values are in the range of 1.30 in areas such as Ezeiza, to 3.5 in Barrio Norte in the federal capital. Changes in the delimitation of each zone and the values of the coefficient Z corresponding to each one must be approved by the regulatory agency. These changes should equal total billing for the period immediately preceding the changes.

AC: Adjustment coefficient (Article 19)

This coefficient is a scale parameter adjusted to maintain the concession holder's revenue level. The initial value of this coefficient was determined from the bidding procedure, in which bidders were required to set the required value of AC. The bidder submitting the lowest value would be selected *ceteris paribus*.

The initial value of the coefficient was 0.731. After rate revision to reflect investments, the value of the coefficient AC was set at 0.830 in accordance with ETOSS Record No. 10.858-94.

Metered charge system

For buildings fitted with a water meter, water service charges use a two-pronged rate which incorporates a fixed charge equal to 50 per cent of the BBR, and a variable charge per cubic metre, expressed with the following formula:

$$FC = 0.5 * BBR$$
$$VC = P * AC * m^3$$

Where P represents the price in $/m^3, AC is the adjustment coefficient. The price for water and sewerage is 0.66 $/ m^3 and for water it is 0.30 $/ m^3.

For residential buildings, the fixed charge includes free consumption of 30m^3. For joint ownership buildings (*edificios de propiedad horizonta*), free consumption is equivalent to total free consumption of each unit in the building.

On an interim basis, during the first two years of the concession, the rate applicable to residential buildings in category RI will include a fixed charge equal to 0.65 * BBR with 42m^3 of free consumption. From the beginning of year three until the end of year five, free consumption will be established using the zone coefficient.

Infrastructure charge

The cost of the household system constructed by the concession holder and the overall supply of new water and sewerage connections in the expansion area may be financed with an infrastructure charge. As this is an alternative system to those provided in the approved expansion plans, it must be authorised in each individual case by the application authority prior to implementation.

The average total reference value for the infrastructure charge was established at the time of the privatisation. For water it was $325 and for sewerage $460. In a resolution issued by ETOSS which modified the value of AC, the reference values for the infrastructure charge were increased to $450 for water and $670 for sewerage.

As stipulated in the Concession Agreement, the infrastructure charge should provide users with two-year financing in equal and consecutive bi-monthly payments. Services should be billed on a combined basis and be clearly itemised.

Connection Charges for Water and Sewerage ($)

Water	
Diameter	Range of Charges
0.013-0032 m	135-200
0.033-0.050 m	175-210
0.051-0.075 m	255-300
Over 0.075 m	340-400

Sewerage	
	Range of Charges
Total	200-225

NOTES

1. BAMA is composed of the City of Buenos Aires (CBA) and many of the municipalities of Greater Buenos Aires (GBA).
2. The firm may request expansion of the covered area to zones not included in the original agreement, subject to authorisation from the regulatory authority.
3. International assistance funds managed by NGOs, multilateral assistance funds, provincial funds, compensation of municipal rates, labour provided by communities.
4. The first layer is profoundly polluted and can be considered unusable. The third layer has a high saline content and has only been tapped for uses where salinity levels are unimportant.
5. This does not include the analysis of the third ring, as it is outside the concession area. Only BAMA townships with major rural components are included.
6. With the assumption that there are no costs associated with the greater absorption of public jobs.
7. While these are 1997 values, the ratio is representative of that used in 1980 when the services were separated.
8. In this case, the equity criterion was replaced with efficiency, reflecting the fact that there is no consumption on undeveloped land.
9. See section 4.6 for a more complete discussion of the elimination of the infrastructure charge in Buenos Aires and its replacement with a universal service charge (SU).
10. The exact ratio for each section is presented in the description of the system for Aguas Cordobesas.
11. Even in cases where the population can connect to illegal water services, the latter does not guarantee a constant supply, adequate pressure, or minimum bacteriological quality.
12. Of course, this is reflected in final costs of treatment, as there is a difference in treating an effluent with a view to water quality compatible with use to protect aquatic life (the most stringent limits under the regulations) and treatment to achieve quality compatible with water for "recreational or consumptive" use.

13. This may be due more to low levels of confidence and credibility in the relevant authorities than to reduced demand *per se*.
14. Following the approval of the CSP, the concept was expanded, and the concession holder is now required to intercept rain and channel it to plants for treatment.
15. While there are arguments of efficiency in support of a scheme of this type, such a scheme can be highly inequitable. This inequity increases if we also consider that users did not pay for existing networks but, given the deficits run by state enterprises that developed the existing network, there was a substantial contribution from the population at large in the form of taxes (explicit and inflationary) to finance the systems.
16. Poor consumers are charged $4 every two months.
17. This financing method is known in some systems as a roll-in mechanism for new investments. In Argentina a similar financing scheme is applied to all customers for specific investments in the gas distribution systems.
18. The use of direct subsidies was proposed by the World Bank for the Buenos Aires concession (World Bank 1995). This suggested direct payment to the concession holder of a fixed rate to subsidise a basic consumption block for each underprivileged consumer. It also suggested avoiding the use of cross subsidies.
19. Programme based on a Co-operation Framework Agreement between Aguas Argentines S.A. and IIED-AL (1997-98).
20. Personal communication, Ana Sosa, ABS Consultores.
21. Conducted in connection with the project "Optimización de la Programación y la Expansión. Demanda Socioambiental de los Servicios" (Optimising programming and expansion. Socio-environmental demand for services (AA/IIED-AL 1996-97).
22. Aguas Argentinas S.A. developed an educational workshop on the use of water, for children between eight and 12 years of age in the municipal schools. Approximately 500,000 children have participated in these workshops.
23. In the case of Barrio San Jorge (San Fernando), owing to the shallowness of the phreatic layer, a specific technological solution was used. This is known as the "decanted effluents" or "sediment-free" system, and it was initially designed for small communities in Australia and the US, and from the late 1980s was adapted successfully to rural and urban communities in Uruguay by MEVIR (movement to eradicate unhealthy housing).
24. IIED-AL's involvement consisted of helping the concessionaire to conduct an interaction process with the local authorities and low-income communities.
25. The annual rate of change was calculated using a cumulative rate of the median year in each period.

5. Private Sector Participation in Water Supply and Sanitation: Realising Social and Environmental Objectives in Manila

Cristina C. David[1]

5.1 INTRODUCTION

In 1997, Metro Manila Water and Sewerage System (MWSS), the government corporation responsible for water supply and sewerage disposal in the greater Metro Manila area, was successfully privatised. The policy decision to privatise was motivated by MWSS's failure under public management to provide adequate water supply and sewerage services to the largest urban centre in the Philippines, and the desire to end government subsidies to its operations.

Inefficiencies in MWSS operations have been widely documented (Binnie/Thames Water/TGGI Engineers 1996). It is also widely considered that these inefficiencies would have been extremely difficult to correct under the same institutional framework and political realities. However, the public good nature of water and its basic need character, economies of scale and externalities in its production, distribution, and consumption, mean that private sector management of urban water resources cannot be expected to achieve economic efficiency, social equity, and environmental sustainability without appropriate contractual arrangements, strong economic and environmental regulations, and other government interventions. These concerns are particularly critical with regard to MWSS because of the Government's weak capacity for enforcing environmental regulations in the face of severe water pollution problems and rapid groundwater depletion within the service area, the bureaucratic constraints encountered in developing a strong economic regulatory office, and the large number of urban poor often bypassed in the allocation of water and sewerage services.

The post-privatisation period of just over a year is too short to evaluate the full impact of privatisation. However, changes in water tariffs and operational costs, analysis of the provisions of the Concession Agreement, and review of the underlying assumptions related to water supply and demand, and other factors enable us to derive some preliminary assessments and explore its potential impact on efficiency, welfare of the poor, and environmental objectives. Moreover, it is important to identify at an early stage potential problems that may be encountered in fulfilling the service obligations and realising the societal objectives of the MWSS privatisation.

The first section of this chapter presents an overview of the institutional and physical characteristics of the urban water supply and sewerage services in the MWSS service area. In the second section, the nature of PSP, contractual and institutional arrangements, and the bidding process are described. The third section examines the general issues and concerns arising from the Concession Agreement that may significantly affect privatisation's impact on efficiency, on the welfare of the poor, and on environmental objectives as analysed in the subsequent two sections. The last section concludes with some policy implications related to overall water resource management, and specifically to the operations of the various components under the present privatised institutional structure of MWSS.

5.2 CHARACTERISTICS OF THE WATER SECTOR

Institutional Structure

The Philippine Water Code (PD 1067) approved in 1976 is considered to be an adequate overall legal framework for the efficient, equitable, and sustainable management of the country's water resources (World Bank 1993a). But the regulatory and institutional frameworks governing the water sector are generally believed to be weak and fragmented. The National Water Resources Board (NWRB), which has overall responsibility for water resource management (i.e., the control, supervision, and regulation of the utilisation, exploitation, development, and protection of the water resources), does not have sufficient authority nor financial resources to perform these functions effectively.

Provision of water-related services and many regulatory functions are carried out by other government agencies. For urban areas outside the MWSS service area, provision of piped water connection is mostly undertaken by water districts which are granted credit subsidies by the Local Water Utilities Administration (LWUA). Local Government Units (LGUs) also manage a small number of water utilities. Since the passage of the Local Government Code (RA 7160) in 1991, responsibility for funding the construction of shallow wells and deepwells for communal use for low income households in their respective political jurisdictions was transferred from the Department of Public Works and Highways (DPWH) to the LGUs. Water use in national irrigation systems is the responsibility of the National Irrigation Administration (NIA), while the National Power Corporation (NAPOCOR) is in charge of the hydro-electric use of water.

Sewerage development is much less organised than water supply because of the limited investment in sewerage. Among the government water utility firms, only MWSS has a mandate for the construction, operation, and maintenance of sanitary sewers and sewage treatment facilities for its service area, as water districts deal only with water supply. The DPWH constructs and maintains storm sewers and drains in Metro Manila, whereas in other urban areas, the LGUs take responsibility for the construction and maintenance of such facilities.

Regulations in the water sector relate to its quantity, quality, and price. The NWRB regulates the use of surface and groundwater resources through its responsibility for granting rights and permits for water abstraction. This regulatory power is shared with the MWSS and water districts as they have the monopoly franchise to provide water services in their respective service areas. In practice, permits for the establishment of private waterworks and individual groundwater usage have been granted quite liberally. This is mainly because the water supply and pipe distribution networks of these franchise holders have not been sufficient to meet total water demand. Moreover, the regulation and monitoring of groundwater abstraction particularly by industrial and commercial establishments and households have been extremely weak. In fact, less than 15 per cent of groundwater users are believed to be registered at the NWRB.

Responsibility for the regulation of water quality is divided between the Department of Health (DOH), for drinking water, and the Department of Environment and Natural Resources (DENR), for the regulation of sewage discharges and industrial effluents. The DENR is also generally responsible for the protection of watersheds, with the exception of a few that are assigned to the NIA and NAPOCOR.

Economic regulation has historically been implicit in the MWSS Charter by the provision limiting the rate of return on book value of assets to 12 per cent. Although, theoretically, the MWSS Board decides on the water tariff subject to the rate of return cap; in practice the price of MWSS water has been politically determined, and ultimately decided upon by the country's president. For the water districts, the LWUA performs the economic regulatory function, while the NWRB is the agency responsible for private waterworks.

Physical Characteristics

The MWSS service area consists of all seven cities and ten municipalities of the National Capital Region (NCR), five municipalities and a city in the province of Cavite, and all 14 of the municipalities of Rizal Province. It covers a geographic area of 2,100 km^2, and in 1995 contained a total of 2.4 million households comprising more than 11 million people.

The bulk (97 per cent) of the MWSS water originates from the Angat Dam (reservoir) and only 3 per cent is from groundwater pumping. The multi-purpose Angat Dam is located to the north, in the adjacent province of Bulacan. It is about 40 km away from the Balara and La Mesa Dam treatment plants in Quezon City from where treated water is distributed through the pipe distribution system. Use of Angat Dam water is shared among the NIA, MWSS, and NAPOCOR. The latter is responsible for operation and maintenance of the facilities, including the Angat watershed.

The NIA has historically had a water right of 36m^3 per second, whereas the water right granted to MWSS has been increasing over time to the current level of 28.8 m^3 per second. Furthermore, the law provides that in the event of a drought, urban water use takes precedence over irrigation and other uses. While the generation of electricity is not a consumptive use of water, the infrastructure was designed such that the capacity to generate electricity declines when water use is shifted away from irrigation.[2] In recent years as drought episodes have become more frequent, the MWSS has withdrawn more than its water right at an average of 32 m^3 per second. Indeed, with the recent severe drought due to the *El Niño* phenomenon, no water was released for irrigation for most of 1998 to ensure Metro Manila's water supply.[3] Despite these moves, MWSS water supply dropped by 20 to 30 per cent during this period.

Water Service

The MWSS piped water connection is estimated to reach only about two-thirds of households within its coverage area (McIntosh and Yñiguez 1997). Its service is generally characterised by low water pressure and intermittent supply, averaging only 16 hours a day. Moreover, MWSS has had the highest rate of non-revenue water among the main cities in Asia, amounting to almost 60 per cent of water production. In comparison, the average rate of non-revenue water among developing countries is 20 to 30 per cent and in Singapore is only about 7 per cent (one of the lowest worldwide). Prior to privatisation, MWSS had nine employees per 1,000 connections, again one of the highest staffing levels relative to other Asian countries (the ratio in Bangkok is only 4.6, Jakarta 7.7, Singapore 2.0, and Kuala Lumpur 1.1).

As a consequence of these shortcomings, a major part of the population and many of the industrial and commercial establishments have had to rely on private waterworks, privately-owned wells, and private water markets. A 1990 groundwater study by JICA (1992) reported that about 40 per cent of total water use and as much as 80 per cent of water use among industrial establishments was supplied from groundwater through private water systems and privately-owned wells (Table 5.1). Moreover, since the sales of MWSS water increased at an annual rate of only 1.3 per cent between 1990-1996 while population in the service area increased at a rate of 4 per cent during the same period, the proportion of water supplied by private wells must have increased significantly, to at least 50 per cent.

Table 5.1: Estimated Water Consumption by Type of User and Source of Water Supply in the MWSS Service Area, 1990 (mld)

	Households	Industry	Commercial	Total (1)	(2)
MWSS	785[a]	75[a]	304[a]	1,163[a] (58)[c]	1,744[b] (67)
% of MWSS	68	6	26		
% of users	69	19	76		
Private wells	379	355	107	841 (42)	841 (33)
% of PW	45	42	13		
% of users	31	81	24		
Total	1,164 (58)	429 (21)	411 (21)	2,004 (100)	2,585 (100)

Notes
a Refers to billed water only.
b Refers to billed water plus estimated non-revenue water consumed through illegal connections and tampered meters. The latter is assumed to be 20 per cent of water production, based on findings of David and Inocencio (1996).
c Figures in parentheses are percentage share of total

Source: JICA (1992)

Overpumping has been depleting groundwater resources since the 1970s and water tables are reportedly falling at rates of six to 12 metres annually in some parts of the service area (Haman 1996, NHRC 1993). As a result, pumping costs are rising and saltwater intrusion and land subsidence have been observed, particularly along the coastal areas. Yet, there has been no effective monitoring, nor regulation of groundwater abstraction. Less than 15 per cent of deepwells are currently registered at the NWRB, and pumping charges are minimal. With the severe water shortage in the Angat Dam caused by the recent *El Niño* phenomenon, the rate of groundwater abstraction has further increased among conjunctive users of MWSS water and privately-owned wells, and as construction of new wells has accelerated. Indeed, the national government granted about 100 million pesos (₱) for the construction of new deepwells which will be integrated into the MWSS water system. LGUs, together with senators and congressmen, have increased the budgetary allocation for the construction of new wells in depressed areas (David and Inocencio 1999).

A 1995 survey of households indicated that because of inadequate coverage and widespread rationing of MWSS water supply, 23 per cent of households in Metro Manila have had to rely on vended water (Table 5.2). Ironically, the major share of water sold through vendors is actually MWSS water, a part of the high non-revenue water caused by meter tampering and illegal connections. Thus, despite government subsidies and the relatively low and progressive water price structure of the MWSS, poor households that are mostly dependent on water vendors end up paying water prices that are as much as ten times higher than high income households that generally have access to piped water connections. Table 5.3 demonstrates more clearly that the average price of water declines as household income increases.

The actual cost of water to many households is often even higher than reported because of the additional costs incurred by the use of storage tanks and booster pumps, the time spent queuing and fetching, and the general inconvenience of dealing with MWSS water rationing and vended water.

Households relying on private waterworks also absorb the initial capital cost of the water system through higher land prices and any further capital expenditure incurred after the waterworks are handed over to the homeowners' association.

Sewerage Service

The MWSS sewerage service is even more limited than its water supply, covering less than 7 per cent of households in the service area. The existing sewerage facilities of MWSS are confined to a few areas in the city of Manila and parts of Makati. Most households and firms utilise privately-owned septic tanks or common septic tanks, while those who live in slum areas without public sewers or drains rely on rudimentary latrines without formal drainage facilities.

Table 5.2: Average Cost of Water and Distribution of Households by Source of Water, Metro Manila, 1995

Source	% of Households	Average Cost (₱/cu m)	Average Monthly Income (₱/capita)	% of Water Bill to Income
MWSS				
(w/o sewer)	51	5.5	2,887	2.0
(w/sewer)	6	8.5	5,648	1.5
Private waterworks	5	7.9	7,249	1.9
Individual tubewell	2	n.a.	5,031	n.a.
Public faucets	1	22-44	n.a.	
Water vendors	23			
MWSS water	19			
Pick-up		30.4	1,168	4.2
Hose (container)		48.3	1,223	6.2
Hose (fixed charge)		21.8	1,325	2.7
Delivered		71.9	1,359	11.9
GW water	4			
Pick-up		40.2	854	5.7
Hose (container)		44.0	2,500	4.8
Hose (fixed charge)		58.9	2,245	3.8
Delivered		62.3	1,850	4.3
Combinations	12			

Source: David and Inocencio (1996), based on survey of 500 households in Metro Manila

Table 5.3: Average Cost of Water by Income Class in Metro Manila, 1995

Income Class	Average Cost * (₱/m³)	% of Water Bill to Income
Under ₱30,000	36.4	8.2
₱ 30,000-39,999	15.9	4.4
₱ 40,000-59,999	15.9	4.2
₱ 60,000-99,999	15.9	2.9
₱ 100,000-149,999	13.9	2.2
₱ 150,000-199,999	9.2	1.6
₱ 200,000-249,999	5.9	1.4
₱ 250,000-499,999	8.0	0.8
₱ 500,000-749,999	6.0	0.8
₱ 750,000-999,999	9.3	0.8
₱ 1,000,000 and over	7.1	0.6

Note * These prices represent the average cost of water from various sources

Source: David and Inocencio (1996)

5.3   MWSS PRIVATISATION

The enactment of the Water Crisis Act (R.A. No.8041) in late 1995, which established the legal basis for the privatisation of MWSS, reflected the Ramos government's commitment to PSP and belief that it would be the most viable approach towards improving the efficiency of MWSS operations, raising financial resources for investments, and ending government subsidies. A new administrator was then appointed whose specific responsibility and whose own personal interest was to privatise MWSS. The International Finance Corporation (IFC) was commissioned to provide technical assistance in developing the process of privatisation, organising the relevant data and its analysis, designing the contractual arrangements and ensuring transparency in the bidding procedures.

Nature of Private Sector Participation

The form of PSP chosen was a 25-year concession agreement, which transfers to a private contractor overall responsibility for operation and maintenance of, and investment in, the water supply and sewerage system. It was also decided to divide the MWSS service area into West and East Zones and grant concessions to two different private companies in order to promote competition and generate yardstick information for more effective regulation (see Figure 5.1 indicating the boundary between zones). The West Zone accounts for about 60 per cent of the population and of water connections in the service area and is also more densely populated. To its west, it is bound by the coastal area of Manila Bay, where groundwater depletion has already lowered water tables, increasing pumping costs and causing saline water intrusion. Because of an older pipe distribution network, the West Zone is characterised by a higher rate of non-revenue water (estimated to be 60 to 70 per cent in comparison to 50 to 55 per cent for the East Zone).

It was also deemed desirable from the perspective of ensuring a stronger financial resource base and technical capability to have foreign private participation, though, as the law stipulates, Filipinos must own at least 60 per cent of the equity. Other requirements relating to relevant experience and financial capability of both local and foreign partners were also imposed for pre-qualification in the bidding process to ensure a competent field of bidders.[4]

Under a privatised MWSS, therefore, the operations, maintenance, and investment for water, sewerage, and sanitation services became the responsibility of the two private concessionaires for the West and East Zones, respectively. The operations of commonly used facilities upstream from the service areas are also undertaken by both concessionaires as a joint venture.

A residual MWSS and its Board have been retained to: facilitate the exercise by the concessionaire of its agency powers; carry out accounting and notification functions; administer domestic and foreign loans related to the existing projects; and manage retained assets including the ongoing development and eventual operation of the Umiray-Angat Transbasin Project (UATP)[5] and other large-scale water supply expansion projects. The manpower complement of the residual MWSS is currently 104, including the members of the Board, an Administrator and Deputy Administrator and staff members for three departments (Administration and Finance, Engineering and Project Management Office, and Asset Management and General Services).

Private Firms and Public Water

Figure 5.1: Metro Manila Water Supply System Concession Service Area

Source: Metropolitan Waterworks and Sewerage System, Engineering Department (1997)

In addition to the residual MWSS, a Regulatory Office (RO) has been established to monitor and enforce compliance by the concessionaires of the contractual obligations under the Concession Agreement, implement rate adjustments, arrange for public dissemination of relevant information, respond to complaints against concessionaires, and prosecute or defend cases before the Appeals Panel. There are approximately 60 employees in the RO, headed by a Chief Regulator and four Regulators (for technical, financial, and customer service regulations, and for administration and legal matters).

Prior to the privatisation, MWSS took a number of steps to facilitate the process, ease problems associated with the transition, and minimise political opposition. Since MWSS was overstaffed (and the strongest opposition to PSP stemmed from its labour force), an early retirement programme was instituted in August 1996 which reduced the number of employees by a third, from 7,500 to 5,200. The average water tariff was also raised from ₱6.43/m^3 to ₱8.78/m^3, primarily to comply with the loan covenant of an ADB-funded project. Although perhaps unintended, the higher average tariff also increased the probability that privatisation would lead to a lower average price of water, and thus make the shift to private sector management more politically acceptable to the public.

Concession Agreement

The Concession Agreement specifies: transitional arrangements; the service, financial, and other obligations of the concessionaires; the obligations of MWSS, including its residual functions, and those of the new Regulatory Office; provision for water charges, rate adjustments and dispute resolution; and other contract conditions. The transitional arrangements relate to transfer of employees, liabilities/revenues, accounts receivable, facilities, existing projects, cash and marketable securities. More than a year after privatisation, the shift from public to private sector management of MWSS - which involved, among other things, organisational restructuring, a reduction of the labour force, and resolution of the issue of connection charges - was implemented without any major difficulties.[6]

Obligations of concessionaires

In terms of service obligations, the concessionaires are required to: expand coverage of water supply, sewerage and sanitation services; provide 24-hour water supply to all connections not later than June 2000 (and substitute alternative supplies at standard rates if source is interrupted for more than 24 hours); by 2007 maintain water pressure at 16 pounds per square inch (psi) for all connections; and meet the national health and environmental standards on quality of drinking water, wastewater discharge, and industrial effluents.

Tables 5.4 and 5.5 report the water supply coverage targets by municipality in the West and East Zones, respectively, every five years from 2001 to 2021. On the whole, the concessionaires are expected to increase the proportion of the population with access to water supply in the coverage area from an initial 67 per cent, to 85 per cent by 2001 and to 96 per cent by 2006 and beyond. The targets are somewhat lower for the East Zone and for municipalities in both zones where the population is more geographically dispersed. By the year 2001, the most heavily populated inner cities of Metro Manila - Manila, Pasay, Quezon, Caloocan, Mandaluyong, San Juan - as well as Cavite City, are expected to be fully covered. By 2006, households in almost all of the cities in the National Capital Region except for Las Piñas, Muntinlupa, as well as three municipalities in Cavite and two in Rizal, should be fully covered.

The concessionaires are required to cover households (in depressed areas who typically do not own their dwellings and may actually be squatters) who may not be able to afford individual connection fees (or where the cost of connection relative to expected revenue may be too high) by establishing public standpipes at a ratio of one per 475 people.

It should be emphasised that the coverage targets on water supply refer to the population less those who already have piped water connections from a source other than the MWSS system.[7] Hence, those obtaining water from their own deepwells or from private waterworks located in areas where the MWSS water service is unreliable, and who are not reached by the distribution network at the time of their establishment, are not covered by the service obligation. In addition, the Agreement does not specify whether or not the coverage includes commercial and industrial establishments. However, the fact that the coverage is expressed in terms of the population may be interpreted to mean that only household or domestic demand for water is considered and not commercial and industrial demand. The exceptions in coverage constitute a major proportion of total water demand. As stated earlier, at least 40 per cent of total water use is estimated to be sourced through groundwater mostly from private deepwells and waterworks (JICA 1992).

The cost of increasing raw water supply, which is required to meet the water service obligations during the first ten years of the concession period, is expected to be financed through investments made by the concessionaires. These investments are to be made, directly, through reducing non-revenue water and rehabilitating old wells and developing new wells and, indirectly, through the concession fee payments used to amortise debts arising from existing water supply expansion projects, including the UATP. The bidders for the West Zone were also advised that an additional 300 million litres per day (mld) of bulk water would be made available by the end of 1999 at no cost to the concessionaire through a Build-Operate-Transfer (BOT) project that will treat Laguna Lake water. It is unlikely that MWSS will absorb the cost of producing such treated water, and thus its cost will have to be passed on to consumers through higher tariffs.

Beyond the tenth year, the contract implicitly recognises the need for another major source of bulk water from surface sources, specifically the Laiban Dam project,[8] in order to meet performance targets. It also stipulates that the cost of such investment will be deemed zero for the concessionaire; in other words, the cost will eventually be passed on to consumers.

Table 5.4: Water Supply Coverage Targets in the West Zone (%) [a]

City/Municipality	2001	2006	2011	2016	2021
NCR					
Manila[a]	100	100	100	100	100
Pasay	100	100	100	100	100
Caloocan	100	100	100	100	100
Las Piñas	58	91	93	95	98
Malabon	84	100	100	100	100
Valenzuela	84	100	100	100	99
Muntinlupa	53	86	88	90	95
Navotas	92	100	100	100	100
Parañaque	76	100	100	100	100
Cavite					
Cavite City	100	100	100	100	100
Bacoor	58	90	92	93	95
Imus	36	61	63	65	72
Kawit	84	100	100	100	100
Noveleta	60	100	100	100	100
Rosario	42	90	90	90	90
Total area[b]	**87**	**97**	**97**	**98**	**98**

Notes

a Expressed as a percentage of the total population in the designated city or municipality at the time of the target (excluding users who are connected to a piped source of water other than from the MWSS system).

b The concessionaire (West) shall also be responsible for meeting the new water supply coverage targets (but not the corresponding sewerage targets), in the percentages set out in Table 5.5 as it appears in the other operator's (East) Concession Agreement, for parts of the following cities or municipalities in the East Zone: Quezon City, San Mateo, Makati, Marikina and Rodriguez.

Source: MWSS (1997a)

Table 5.5: Water Supply Coverage Targets in the East Zone (%)[a]

City/Municipality	2001	2006	2011	2016	2021
NCR					
Mandaluyong	100	100	100	100	100
Makati[b]	92	100	100	100	100
Marikina[b]	92	100	100	100	100
Quezon City[b]	100	100	100	100	100
Pasig	92	100	100	100	100
Pateros	84	100	100	100	100
San Juan	96	100	100	100	100
Taguig	44	100	100	100	100
Rizal					
Angono	51	96	98	100	100
Antipolo	78	95	95	95	97
Baras	34	51	53	55	58
Binangonan	40	81	83	85	87
Cainta	64	80	77	75	79
Cardona	34	51	53	55	58
Jala-Jala	34	51	53	55	58
Morong	34	51	53	55	58
Pililla	34	51	53	55	58
Rodriguez	83	95	95	95	98
San Mateo	84	100	100	100	100
Tanay	39	75	75	75	76
Taytay	92	100	100	100	100
Teresa	52	60	60	60	61
Total area [c]	**77**	**94**	**94**	**94**	**95**

Notes

a Expressed as a percentage of the total population in the designated city of municipality at the time of the target (excluding users who are connected to a piped source of water other than from the MWSS system).

b A portion of this municipality is covered by the West Zone.

c The concessionaire (East) shall also be responsible for meeting the new water supply coverage targets (but not the corresponding sewerage targets), in the percentages set out in the other operator's (West) Concession Agreement, for part of Manila in the West service area.

Source: MWSS (1997a)

Sewerage and sanitation

The coverage targets for sewerage and sanitation services are limited to households connected to the MWSS water system. Tables 5.6 and 5.7 show the coverage targets separately for sewer connection and sanitation services by municipality in the West and East Zones, respectively. On the whole, coverage for sewer connection is scheduled to increase slowly from 7 per cent at the beginning of the concession period to 14 per cent in 2001 and 18 per cent in 2006, rising to 62 per cent by 2021 as the development of sewerage infrastructure is completed. In the meantime, sanitation services, defined as the de-sludging of septic tanks every five to seven years, is likely to be the most common means of addressing domestic sewage problems. Target coverage of sanitation services is scheduled to decrease over time from about 41 per cent in 2001 to 24 per cent by 2021.

Financial obligations

The financial obligations of the concessionaires relate to the size of equity investments, the performance bond and the various fees designed to free the national government from subsidising MWSS as it had done historically. In terms of equity investments, each of the local and international partners are required to maintain an equity share of 20 per cent for the first five years and 10 per cent thereafter. The initial cash equity investments will be ₱3 billion ($100 million) for the West Zone and ₱2 billion ($67 million) for the East Zone. To be renewed annually, a performance bond of $120 million for the West Zone and $70 million for the East Zone must be maintained during the initial ten years, after which the performance bond declines for each successive rebasing date. The penalty for non-compliance with the Concession Agreement by the concessionaire will be deducted automatically from the performance bond.

Upon the takeover of the MWSS operations, a commencement fee of US$5 million was collected from each concessionaire. This revenue was used to pay for the costs of the privatisation process, including the technical assistance contract with the IFC.

Concession fees are to be paid to cover the amortization payments of the local and foreign debts of the MWSS, and the costs of the operations of the residual MWSS and its Regulatory Office. For the latter, each concessionaire will contribute ₱100 million of a total of ₱200 million which will be distributed equally between the Regulatory Office and the residual MWSS.

The West Zone was charged substantially more (90 per cent) of the total amortization payments than the East Zone (10 per cent) as concession fees.[9] Tables 1 and 2 in Annex 1 to this case study show the concession fees to be paid by the West Zone and East Zone, respectively, which decline sharply over time as existing debts are paid off.

Table 5.6: Sewer and Sanitation Coverage Targets in the West Zone (%)[a]

City/ Municipality	Sewer [b]					Sanitation [c]				
	2001	2006	2011	2016	2021	2001	2006	2011	2016	2021
NCR										
Manila	55	71	77	83	91	9	9	9	9	9
Pasay	0	0	0	16	95	73	68	66	47	0
Quezon City	0	0	0	0	54	41	37	38	97	45
Caloocan	3	2	2	32	79	30	61	47	42	21
Las Piñas	0	0	0	0	50	46	57	50	41	27
Malabon	2	2	2	38	94	7	42	39	35	6
Muntinlupa	0	44	57	54	61	27	36	31	26	24
Navotas	3	3	3	36	90	14	65	60	54	10
Parañaque	0	0	0	0	52	53	59	53	46	42
Valenzuela	0	0	0	24	59	67	90	80	68	36
Cavite										
Cavite	0	0	0	0	0	100	89	84	91	86
Bacoor	0	0	0	0	0	52	67	60	56	50
Imus	0	0	0	0	0	11	15	15	24	24
Kawit	0	0	0	0	0	67	68	61	52	47
Noveleta	0	0	0	0	0	28	41	39	35	33
Rosario	0	0	0	0	0	14	25	23	20	18
Total	16	20	21	31	66	43	46	43	39	27

Notes

a Expressed as a percentage of the total population in the designated city or municipality connected to the concessionaire's water system at the time of the target.

b The concessionaire will also be responsible for meeting sewer coverage targets in the part of the City of Manila covered by the other operator unless obstructed from doing so by a natural waterway.

c The concessionaire shall also be responsible for meeting sanitation coverage targets (in the percentages set out in the other operator's Concession Agreement) for parts of the municipalities of Makati, San Mateo, Marikina, and Rodriguez in the East Zone.

Source: MWSS (1997a)

Table 5.7: Sewer and Sanitation Coverage Targets in the East Zone (%)[a]

City/ Municipality	Sewer [b]					Sanitation [c]				
	2001	2006	2011	2016	2021	2001	2006	2011	2016	2021
NCR										
Quezon City	0	0	83	87	98	24	21	16	12	2
Mandaluyong	0	0	100	100	100	0	0	0	0	0
Makati	22	52	100	100	100	0	0	0	0	0
Marikina	0	0	0	0	0	63	79	73	64	60
Pasig	0	41	68	68	68	83	58	32	27	25
Pateros	0	60	100	100	99	0	0	0	0	0
San Juan	0	0	100	100	100	0	0	0	0	0
Taguig	0	52	75	84	100	0	0	0	0	0
Rizal										
Angono	0	0	0	0	0	19	30	49	44	41
Antipolo	0	0	0	0	0	57	53	63	50	44
Baras	0	0	0	0	0	0	0	0	0	0
Binangonan	0	0	0	0	0	12	21	26	23	22
Cainta	0	0	0	0	14	38	40	34	28	27
Cardona	0	0	0	0	0	10	13	12	10	10
Jala-Jala	0	0	0	0	0	0	0	0	0	0
Morong	0	0	0	0	0	0	0	0	0	0
Pililla	0	0	0	0	0	0	0	0	0	0
Rodriguez	0	0	0	0	0	0	0	0	0	0
San Mateo	0	0	0	0	0	66	65	58	49	44
Tanay	0	0	0	0	0	0	0	0	0	0
Taytay	0	0	0	0	15	82	78	70	60	54
Teresa	0	0	0	0	0	25	25	23	21	20
Total	**3**	**16**	**51**	**52**	**55**	**38**	**32**	**27**	**24**	**19**

Notes

a Expressed as a percentage of the total population in the designated city or municipality connected to the concessionaire's water system at the time of the target.

b The Concessionaire will also be responsible for meeting sewer coverage targets specified in Schedule 4 in the part of the cities or municipalities of Makati, San Mateo, Marikina, and Rodriguez covered by the other operator unless obstructed from doing so by a natural waterway.

c The Concessionaire shall also be responsible for meeting sanitation coverage targets as it appears in the other operator's Concession Agreement for the part of the city of Manila in the West Zone.

Source: MWSS (1997a)

Other provisions

Water charges. The average tariffs will initially be set based on the bid price, expressed as the percentage of the current average tariff to which the concessionaire will reduce tariffs. That percentage is to be applied to the current increasing block tariff structure that is higher for commercial and industrial users compared with household consumers. In addition, the concessionaire may apply a Currency Exchange Rate Adjustment (CERA) charge of ₱1 per cubic metre of water consumed and collect a connection charge for water or sewer connection not exceeding ₱3,000 (adjusted for inflation) for distances of less than 25 metres between the connection point and the customer and at a reasonable cost for customers further away.[10]

Rate adjustments. The Agreement provides for water tariff rate adjustments from time to time, subject to the MWSS's charter limitation on its rate of return, which is equal to 12 per cent of the book value of its assets. As will become clear below, that limitation is essentially redundant because the Agreement's effective cap on the concessionaire's rate of return on its own investments is reflected in the Appropriate Discount Rate (ADR). The Agreement stipulates that the ADR should be "in line with the rates of return being allowed from time to time for the operation of long-term infrastructure concession arrangements in other countries having a credit standing similar to that of the Philippines". At least in the first five years of the concession period, the ADR is what is implied by the financial bid price of the concessionaires, which presumably reflects the rate of return they are willing to accept for managing the concessions according to the contract.

There are three bases for rate adjustment: inflation, extraordinary circumstances, and rebasing. Inflationary factors are explicitly stated as grounds for changes in connection charges. In terms of water tariffs, adjustment for inflation is also allowed implicitly by the fact that bidders were given to understand that over the life of the concessions, the rate of inflation would be considered zero. Grounds for extraordinary price adjustments include amendments in the service obligations, changes in the law and other government regulations that affect cash flows, the existence of below-market interest rate financing from any multilateral or bilateral sources, movements in the exchange rate above 2 per cent, erroneous bidding assumptions provided by MWSS prior to the bid, increases in the concession fees, delays in the completion of the UATP, and increases in the operational cost as a result of an uninsured event of *force majeure*. The latter includes among others, war, volcanic eruptions, unusually severe weather conditions, prolonged strikes, and any other event which is not within the reasonable control of the concessionaires.

Whereas inflation and extraordinary circumstances may be allowed as grounds for price adjustment any time after the first year, rate rebasing follows a five-year cycle. The Agreement specifies that from the tenth year or the second rate rebasing date, water tariffs shall be set to allow concessionaires to recover over the concession period, "operating, capital maintenance, and investment expenditures efficiently and prudently incurred; Philippine business taxes and payments corresponding to debt service on the MWSS loans and concessionaire loans incurred to finance such expenditures, and to earn a rate of return (referred to herein as the Appropriate Discount Rate) on these expenditures for the remaining term of the Concession". The Regulatory Office, however, may decide to consider a rebasing adjustment earlier, on the first rebasing date or the fifth anniversary of the concession's commencement date, which is the year 2003.

Taxes. The concessionaires are granted a six-year income tax holiday, a preferential tariff of 3 per cent on capital equipment imports and tax credits on locally fabricated capital equipment until the end of 1997, and exemptions from local government and franchise taxes and value added tax (VAT) on the supply and distribution of water. But a 10 per cent VAT will be applicable for the provision of sewerage and sanitation services.

Bidding Process

Based on pre-qualification criteria, four companies were shortlisted.[11] These companies were required to bid for both the West and East Zones by first submitting their technical bids and plans for achieving the service obligations specified in the contract. After evaluation of the technical bids, which all four companies passed, the next step was the submission of financial bids expressed in terms of the percentage of current average tariffs to which the concessionaire would reduce water tariffs.

Unexpectedly, the Ayala/International Water (AIW) financial bids for both zones were far lower (25 to 30 per cent) than those made by the other companies, which submitted fairly similar bids ranging from 50 to 60 per cent. Since a company may only win one of the concessions, the average tariff in the East Zone won by the Ayala/International Water (now called the Manila Water Company), turned out to be only about half (₱2.32 or $0.09/m^3) that of the West Zone (₱4.97 or $0.19/m^3) which was won by Benpres/Lyonnaise (now Maynilad Water Services Inc.).[12] The bid prices were generally low even by comparison to the earlier price of ₱6.43/m^3, which was raised to ₱8.78/m^3 a few months before the financial bidding. Metro Manila now has the lowest priced water in the country and in the ASEAN region (see Annex 1, Tables 2 and 3) - Bangkok has the next lowest average price with $0.31/m^3 while the highest is Jakarta, with $0.61/m^3 (McIntosh and Yñiguez 1997).

Ironically, the decision to have two separate concessions in order to guard against monopoly profits resulted in a situation where a higher bid price had to be accepted and the price of MWSS water differed significantly between the two zones. It was generally believed that the higher bid prices provided the normal rate of return, while Manila Water's bid was too aggressive and would not be financially viable over the long-term. But the bidding procedure did not specify any minimum financial bid nor any mechanism to prevent wide disparity in water prices in the event that winning bids differed substantially between the two zones. Although the concession fee is much higher in the West Zone, this was supposed to be balanced by the expected lower operational cost per cubic metre due to the higher population density in this service area, and thus similar financial bids were expected for the two zones. Interestingly, the financial bids of each company were similar for the two zones. Such wide tariff differences between the two zones were never intended, and are clearly unfair to the customers.

5.4   GENERAL ISSUES AND CONCERNS

It is obviously too early to evaluate fully the impact of the MWSS privatisation on efficiency, the poor and the environment.   Nonetheless, analysis of the provisions of the Concession Agreement and the underlying technical and business assumptions used in decision-making is useful in deriving preliminary assessments of its potential impact. Whether or not the societal objectives with regard to water resource management in Greater Metro Manila are achieved, however, depends not only on the performance of the privatised MWSS but on the effectiveness of the overall regulatory and policy framework.

The water sector's performance under a privatised MWSS structure so far indicates a number of efficiency gains.   MWSS water service is now delivered at a much lower cost to consumers, especially in the East Zone. The number of employees has fallen by another 20 to 25 per cent, and now averages five employees per 1,000 connections. Meter replacements have been accelerated, repairs to leaks are carried out more promptly, and a variety of measures has been adopted to reduce illegal connections. Monitoring of water quality and wastewater effluents from the sewerage system has become more systematic, and enforcement of environmental standards more effective. Quarterly service performance reports relating to the fulfilment of the concessionaires' obligations are submitted and these are verified by the Regulatory Office.

Improvements in the management of water supply have also been apparent from the more timely and effective response of the various government agencies concerned with the severe drought caused by *El Niño* compared to previous drought episodes: farmers were informed about the lack of irrigation water before the planting season; an orderly rotation of scarce urban water supply was implemented; mobile and stationary tankers were deployed in depressed areas; and public expenditure for shallow wells and deepwells was increased.

The serious shortage of raw water from the Angat Dam (25 to 30 per cent reduction) and sharp devaluation of the peso (about 60 per cent) within the first year of privatisation, however, have had significant negative effects on the concessionaires' net cash flows. By March of 1998, the two concessionaires petitioned for upward rate adjustments, but only a small increase was granted to compensate for the impact of devaluation, as the water shortage due to *El Niño* was considered a recurrent phenomenon that should have been taken into account.[13] Following the rate adjustment, the difference in water rates between the two zones has become greater.

Appeals have been submitted for reconsideration and all parties now expect that a rebasing adjustment will be requested at the first rebasing date or in the fifth year. Although the Government should control possible monopoly profits by the private water concessionaires, pricing policy must be evaluated more broadly as a means for establishing the correct level of incentives so that an adequate water, sewerage, and sanitation service may be provided to all at the minimum cost and at prices consumers are willing to pay.

Analysis of the implications of MWSS privatisation on efficiency, the poor and the environment should not be based solely on MWSS-specific issues, but should also consider the overall regulatory, institutional, and pricing policy frameworks affecting the water sector. There are a number of major reasons why the potentially positive impact of privatisation may be limited.

Pricing Policy

The pricing policy implied by the Concession Agreement and the bidding procedure does not take full account of the opportunity cost of water and the cost of externalities in water production and consumption. The water tariff structure now supposedly reflects the minimum average financial cost of treatment (and production of groundwater) and distribution of water, as well as sewerage and sanitation services. Prior to privatisation, the pricing policy similarly accounted only for the financial cost of MWSS operations, but it also covered the cost of inefficiencies under public sector management.

There is no price charged for the raw water from the Angat Reservoir despite competing use for that water among irrigation, urban use, and electricity generation, as already established in the ADB study (Young et al. 1996). The NWRB does not levy any significant pumping charge, although there are competing uses for groundwater, and depletion has already increased pumping costs and caused saline water intrusion and land subsidence in many parts of the service area. Consequently, such a pricing policy will misallocate water resources in favour of lower valued uses, worsen groundwater depletion, and promote wasteful usage of water.

Politically, the imposition of the appropriate raw water charge for Angat water and pumping charges for groundwater at the time of the MWSS privatisation would have been very timely. A raw water charge for Angat water (or a pumping charge) of as much as $P2/m^3$, based on the 1996 ADB estimate of the economic value of long-term water transfer from irrigation to urban use, could have been imposed without increasing the water tariffs to the consumers.

The IFC, apparently, proposed such a raw water charge for Angat water, but the idea was rejected because it was considered that it would further raise water tariffs, and the very low bid prices were quite unexpected. It was also believed that revenue from such charges would accrue to the general treasury and may not benefit the water sector nor the consumers in the service area. Furthermore, because of the fragmentation of water resource management, the need for stronger regulation of groundwater pumping and for charging the opportunity cost of Angat water through pricing mechanisms was not considered during the MWSS privatisation process. Revenues from such charges could have funded the much needed water-related investments in groundwater recharge, watershed protection, administration costs for more effective water resource planning and management, enforcement of regulations, and possible subsidies for the water-related needs of the poor.

Coverage Targets

It should be noted that the interpretation of water supply coverage targets seems to differ between what is socially desirable and thus should be proposed by the government and what is actually stipulated in the Agreement.

The Agreement stipulates that the water supply coverage target excludes users who are connected to a piped source of water other than from the MWSS system, and says nothing about the targets for the commercial and industrial users of water. This means that a significant number of households and industrial and commercial firms relying fully or partly on their own wells and private waterworks will not have to be supplied with MWSS water. Yet, surface sources of water supply which can only be developed viably on a large scale will have to replace pumping wells if groundwater depletion is to be addressed. Available data also indicates that the full economic cost of groundwater pumping is likely to be greater than the cost of expanding surface water supply through the sectoral reallocation of Angat Dam or the construction of the Laiban Dam (Young et al. 1996, Electrowatt and Renardet 1997). Households and other water users would undoubtedly be willing to pay significantly higher prices for water than the new MWSS prices, as evidenced by the number of consumers with MWSS connections who often use booster pumps and storage tanks, and those depending on privately-owned wells, private waterworks, and vended water.

According to several former and current officials of the MWSS, the targets were intended to cover everyone who wished to receive MWSS water. The exclusion clause was not deemed to be a problem because it was considered to be in the interest of the concessionaires to expand coverage. However, this assumption is only true if the marginal revenue from increasing water supply is greater than the marginal cost, or when the Angat water or other bulk water priced at zero, is plentiful. While reductions in non-revenue water (NRW) from 60 per cent to 30 per cent may be easily accomplished by addressing the problem of illegal connections and water theft, this will simply increase revenues but not water supply. Given the present water rate structure, however, the marginal cost of increasing water supply by further reductions of NRW through pipe rehabilitation or intensifying groundwater pumping is likely to be higher than marginal revenues.

Demand Projections

The concessionaires determined their financial and technical bids for the right to operate the MWSS in return for meeting the service, financial, and other obligations. These bids were made on the basis of information including: projections of water demand; expected raw water supply from the Angat Reservoir and other sources which should be available at no cost to the concessionaires up to the tenth year; the size and quality of the facilities; current sources of water supply of households and firms within the service area, and so forth. A recent review of water demand projections for the MWSS service area suggests that these are generally underestimated because of faulty assumptions, limited data availability, and dearth of empirical analysis and economic estimations of water demand relationships (David 1998). For example, official population projections by the National Statistics Office have proven to be consistently underestimated. There is no reliable information about the commercial and industrial water use which is largely supplied from private wells. Although it is probable that the IFC and the private bidders made their own projections (although they are not published), these are likely to have had the same shortcomings.

Figure 5.2 shows projected surface sources of water supply, sustainable groundwater supply, and more reasonable (i.e. higher) water demand projections carried out by Electrowatt and Renardet (1997) and David et al. (1998) which turned out to be quite similar.[14] If only the sustainable groundwater abstraction of about 500 mld is permitted (100 mld by MWSS and 400 mld for private wells), these figures indicate substantial shortfalls in surface water supply over the long-term, unless demand for water declines (which is only likely if the price of water is raised to cover the full economic cost and/or regulations governing water pollution are strengthened).

In any case, if water demand projections have been underestimated, water supply from surface sources will be quite limited, and water supply performance targets may not be met before the Laiban Dam Project is completed. In cases of water shortage, it is usually the poor who do not have access to the low-priced MWSS water. Even if the narrow coverage targets as strictly defined in the Agreement are met, this would mean increasing reliance on private waterworks and wells, and worsening groundwater depletion because users will not wait for coverage targets to be met.

Figure 5.2: Projections of Water Demand and Supply in the MWSS Service Area, 1990 to 2015

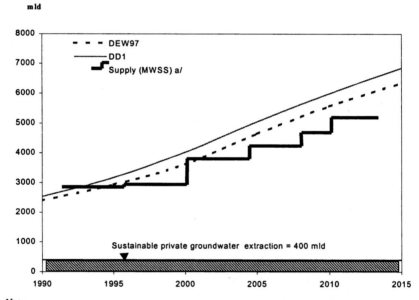

Note

a Projected net supply of MWSS water, i.e., net of unaccounted for water. The base year 1990 estimate is equal to the 1989, 1990, and 1991 average sales plus adjustment for the amount of non-revenue water that is actually consumed but unbilled. The latter is assumed to be one-third of non-revenue water. Non-revenue water is the difference between the actual water production and the actual sales; for 1990 and 1995, NRW are 58% and 56% respectively, while for 2000, it was assumed to be 30% and for 2005-2015, 20%. By 2000, a net supply of 770 mld will be added with the completion of the UATP that will produce 800 mld and the BOT contract to produce 300 mld of treated Laguna Lake water. By 2005, 2009, and 2012, the proposed completion of the Laiban Dam if it starts in 1998 will raise gross supply by 650, 650 and 600 mld, respectively.

Source: David et al. (1998)

Increasing Block Tariff Structure

The MWSS's highly complex increasing block tariff structure has been adopted by the private concessionaires, though the Agreement permits tariff structure adjustments subject to approval by the Regulatory Office (see Annex 1, Tables 1 and 2). There are four categories of user: Residential, Semi-business, Business I, and Business II. Tariffs are lowest for households and highest for large-scale industrial or commercial users of water (Business II). The difference between these two types of user was much greater at the lowest consumption block of the first ten cubic metres (five times) compared to the higher blocks (50 per cent difference).

Residential and semi-business connections have nine blocks, while 33 blocks are detailed for the business connections. Residential connections are characterised by sharply rising block tariffs (four times from lowest to highest), semi-business connections are more gradual (2.5 times), while business connections have relatively flat rates (1.4 times).

The increasing block tariff structure is usually justified as a means of cross-subsidising the poor households, promoting water conservation, and being consistent with marginal cost pricing. There is growing recognition, however, that a highly complicated increasing block structure has serious disadvantages, often defeating its original purpose (Boland and Whittington 2000).

In practice, most of the poor are unable to obtain individual MWSS connections, but instead have to rely on high-cost vended water. Many of the poor also share MWSS water connections or standpipes, or buy water from neighbours with private connections. Moreover, because of the increasing block tariff structure these poor households end up paying water prices at the high end of the structure as total water use exceeds the lower consumption blocks. The initial minimum water consumption block of $10m^3$ per month is also typically higher than water use of the very poor households with individual MWSS connections (estimated at about 3-5 m^3), effectively raising the unit price of water for the poor relative to the middle-income households consuming greater quantities of water.

The much higher water tariffs for commercial and business establishments, coupled with economies of scale in deepwell operations, promote groundwater pumping and thus exacerbate its depletion. There may also be efficiency losses as the marginal cost of increasing water supply from large-scale surface sources may be lower than from deepwells, especially when the full economic cost of groundwater pumping is considered.

The very complex increasing block water tariff structure also makes it more difficult for the operator to project revenues and for the consumer to determine the marginal cost of increasing its water use. The price elasticities of water demand by income class, by type of user, and by size of firm are seldom known, but these parameters are necessary in deriving revenue projections. On the other hand, when the price signals are not transparent, such as in block tariff structure, consumers will not be able to respond properly to them.

5.5 IMPLICATIONS FOR THE POOR

The Government has historically addressed the water needs of the poor in two general ways. For MWSS (and other public water utility firms), an increasing block tariff structure together with higher prices for commercial and industrial firms is adopted as the pricing policy to cross-subsidise poor households. In Metro Manila, household surveys in 1995 and 1998 (David and Inocencio 1996, 1999) indicated that the majority of low-income households do not have individual piped water connections, but rely instead on vended water. Most poor households are not eligible for water connection because either they do not own the title to the land or the public and private owners of the land will not give their permission. Although in the past there has been a black market for obtaining water connections, an ordinary low-income household would not be able to afford its high cost. Many poor households also live in areas outside the pipe distribution network.

MWSS has established standpipes in squatter areas but these are very few. A recent count (ADB 1993) shows the ratio of standpipes to total number of connections to be less than 0.2 per cent. Furthermore, a preliminary assessment suggests that a significant number of these are not operational either because the management has failed to remit collected funds, or water supply is so intermittent that households have not been willing to continue payment.

The other approach has been to subsidise directly the construction of shallow wells and deepwells of varying sizes for common use by communities in depressed areas. Prior to the passage of the Local Government Code in 1991, this was managed by the Department of Public Works and Highways. By 1992, that responsibility was transferred to local governments. Both congressmen and senators have allocated some of their pork barrel funds for that purpose, and in the recent election, political candidates also donated funds for construction of wells to win votes among poor households.

The impact of privatisation on the poor depends largely on the adequacy of MWSS water supply. Provisions in the Agreement that relate to the poor are the retention of the increasing block tariff structure and the establishment of public standpipes for every 475 people within depressed areas with no installation charge. However, with water supply shortages, the poor would tend to receive low priority, especially if the concessionaires were compelled to charge the lowest tariff block in these areas. The Agreement did not make that provision and thus poorer households covered by the performance targets would probably be served through sharing water connections or public standpipes, paying higher prices than middle and higher-income households. The price ultimately paid by poor households will also depend on how the distribution of water from the public standpipe will be managed.

On the ground, a wide variety of formal and informal mechanisms for distributing water from different sources is evolving in the low income areas. Preliminary results in the recently completed survey of low-income households show that only about 20 to 25 per cent of respondents have individual MWSS connections (David and Inocencio 1999). The majority rely on vended water sourced from MWSS connections or pipes and on sharing the water bill from an MWSS connection. Average cost of shared MWSS water is at the higher end of the tariff structure while the price of vended MWSS water ranges from about ₱30/m³ when based on a fixed rate and picked up from the source, up to about ₱200/m³ when MWSS water is delivered by trucks. More common is the practice of vendors selling MWSS water by container at a price of approximately ₱50/m³. The remainder would be evenly divided between those who rely on public pumps or public faucets and vended water from deepwells. The price of water from public pumps/faucets ranges from an average of ₱10/m³ to more than ₱40/m³, depending on the quality of water and the location. Vended deepwell water costs from about ₱100/m³ when picked up or transferred by hose to as high as ₱150/m³ when delivered by trucks.

Comparing the 1995 and 1998 surveys of low-income households, certain trends may be observed (David and Inocencio 1996, 1999). The proportion of household respondents with MWSS connections has decreased, as public pumps and faucets, water vending, and sharing of MWSS connection have become more significant mechanisms for water distribution. It is interesting to note that the price of vended water from both MWSS and deepwell sources has increased significantly despite a reduction in the average official price of MWSS water.

The above trends are to be expected given that MWSS water supply has declined due to the *El Niño* phenomenon, demand for water in general and for other sources of water in particular has increased, and the cost of groundwater pumping has risen. As argued earlier, limited water supply results in the benefits of low-priced MWSS water accruing mostly to the high and middle-income households. While the average price of MWSS water from shared connections is lower than vended water, the increasing block tariff structure has meant that higher income customers pay a lower unit price for water than lower income households who have to share the water bill from a single connection. This is quite unfortunate since sharing of water connections would be one of the more efficient ways of extending access to the MWSS water service to poor households who cannot afford a separate connection. On the other hand, charging the lowest price for water

distributed through shared connections and standpipes would only encourage
fraud and lessen the incentive for the concessionaires to allocate more of the
scarce water supply to low-income areas.

*Figure 5.3.1: Water Tariff Structure for Residential Dwellings before and
after Privatisation (P/m³)*

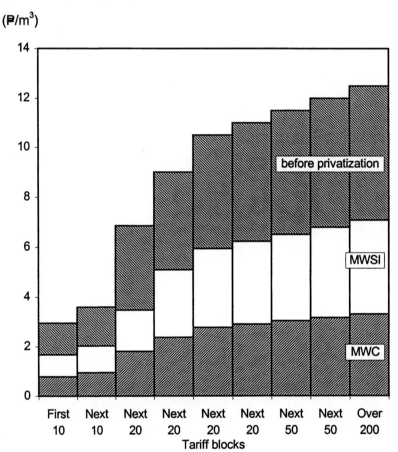

Figure 5.3.2: Water Tariff Structure for Semi-business Dwellings before and after Privatisation (₱/m³)

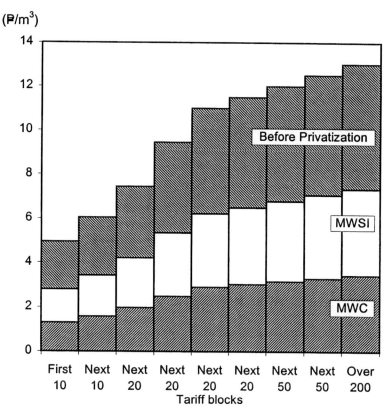

5.6 IMPLICATIONS FOR THE ENVIRONMENT

Health and Water Pollution

Privatisation could potentially make major contributions towards improving health and reducing water pollution. It attempts to internalise the externalities in water consumption through a more organised and expanded effort at dealing with sewerage and sanitation problems and full cost recovery pricing. Stricter enforcement of drinking water and wastewater standards may be expected because the Regulatory Office is adequately funded and dedicated to monitoring and enforcing these standards. Moreover, there is a greater incentive to comply as non-compliance means breaching one of the service obligations. As mentioned earlier, sewerage and sanitation services prior to privatisation were extremely limited, and though wastewater and industrial effluent regulations exist, enforcement has been very weak. Some progress has been made through the recent imposition of an effluent charge instituted by the Laguna Lake Development Authority (LLDA) which covers a large part of the MWSS service area.

It should be stressed, however, that the exclusion of a segment of the population and the commercial and industrial users of water from the water supply and sewerage and sanitation targets leaves a significant part of the population with little government support to ensure good quality drinking water. The Concession Agreement also overlooks the larger part of wastewater from commercial and industrial establishments, self-supplied water users and those dependent on private works who will have no access to a possibly more efficient and less costly means of dealing with wastewater and industrial effluents. Nonetheless, without MWSS privatisation, expansion of sewerage and sanitation services would have been unlikely, especially in depressed or squatter areas which may now be covered by the Concession Agreements. It is worth pointing out that diarrhoea, a waterborne disease, is the principal cause of morbidity and the fourth largest cause of infant mortality in the country (National Statistics Office 1998).

Water Conservation and Groundwater Depletion

The imposition of higher sewerage and sanitation charges that largely take account of the cost of externalities involved in water consumption will increase incentives to save water, at least the opportunity cost of water, among customers covered by such services. To the extent that conjunctive users of privately-owned wells and MWSS water do not have to pay sewerage or sanitation charges from self-supplied water, the incentive to conserve water is in part dissipated.

The price of MWSS water continues to reflect only the financial cost of production and distribution of water and sewerage and sanitation services, and no charges are imposed to cover the opportunity cost of water, the cost of externalities of groundwater pumping, or the external cost of water consumption among self-supplied water users and those dependent on private waterworks. Thus, wasteful usage of water, overpumping of groundwater, and pollution of water bodies will likewise continue to have negative environmental and health consequences.

The failure to impose a raw water charge to cover the opportunity cost of Angat water and a pumping charge for use of groundwater also lessens the incentive for concessionaires to invest in reducing non-revenue water since the net gain would be lower than when these costs are considered. The lack of any significant pumping charge for groundwater use means that groundwater depletion will continue unabated. Severe water shortage in the Angat

Reservoir due to the recent *El Niño* phenomenon has already led to increased investments in deepwells by households, industrial and commercial establishments, national government and local governments throughout the MWSS service area.

Underestimation of demand projections has lessened the perceived severity of surface water shortage and pushed back the scheduling of surface water supply development, specifically the Laiban Dam construction. Such large-scale, long-term water supply expansion projects will not be undertaken by the concessionaires within the contractual tariff rate. Hence, abstraction of groundwater will accelerate as the concessionaires rely on deepwells to meet performance targets and real estate developers, households, industrial and commercial establishments will have to cope with the limited services of the water utility firms.

The treatment of Laguna Lake water to provide 300 mld of water supply to the West Zone will partly alleviate the supply gap. Thus far, however, there has been no study into the environmental implications of such consumptive use of lake water, nor into the financial viability of such an undertaking if pollution of lake water continues unabated.

5.7 CONCLUDING REMARKS

The privatisation of MWSS is an important and positive step towards improving water resource management in greater Metro Manila. Realising the full potential gains from that initial step in the long-term depends critically on the following:

- the ability of the Regulatory Office and the residual MWSS to enforce the contractual agreements (the spirit as well as the letter), to anticipate potential problems arising from possible weaknesses in the contract design and changes in the underlying assumptions, data, and analysis used in developing the contract and the technical and financial bids, and to implement expeditiously the necessary adjustments in the contract and mode of operation.

- the willingness of the Regulatory Office and the residual MWSS to adopt a more integrated and holistic approach in dealing with the inherently interrelated issues of water supply and sewerage planning and operations, demand management, pollution control, and watershed and groundwater protection.

- the Government's ability to undertake the necessary institutional, regulatory, and policy reforms in the water sector to ensure effective coordination of policies and programmes and establish appropriate incentive and control structures for more efficient, equitable, and sustainable management and utilisation of water.

It should be emphasised that the regulation and management of the privatised MWSS structure must be evaluated from the perspective of achieving the overall objective of economic efficiency, social equity, and environmental sustainability. These should not be viewed narrowly from the perspective of enforcing contractual agreements and minimising water prices. Clearly, the adoption of full economic cost pricing policy is a critical step which would involve the imposition of a raw water charge on Angat water, a pumping charge for groundwater abstraction, and a water pollution tax beyond the LLDA coverage area. Government revenues from such charges may be earmarked for water resource management-related activities, including strengthening of the public sector's technical capability for planning, policy analysis and formulation, regulatory design and enforcement. These may involve: improvements in the statistical database on groundwater abstraction and recharge, water quality, streamflow of relevant river systems; conducting analytical studies for more accurate water demand projections and water supply and sewerage planning, and other long-term research on resource management issues; subsidising the cost of water, sewerage, and sanitation service provision to poor households; funding efforts for increasing groundwater recharge and strengthening watershed and groundwater protection; and financing capital and operational costs of water treatment facilities and other methods for rehabilitating polluted water bodies.

To resolve the other issues and concerns raised about the provisions of the Concession Agreement, it is necessary to undertake the following: a review of water demand-supply to take into account the problem of groundwater depletion in order to make the necessary adjustments on water supply development projects and coverage targets; a reduction in the difference in water tariffs between the two zones; simplification of the water tariff rate structure to narrow the wide price difference across types of users and consumption blocks. Finally, it may be necessary to review the appropriate level of sanitation charges, especially if water tariffs increase significantly.

ANNEX 1: DATA TABLES

Table 1: Land Area, Number of Households, Population, and Population Density of the Cities and Municipalities in the MWSS Service Area, 1995

	Area (km²)	Number of Households (000)	Population (000)	Population Density (000/km²)
MWSS Service Area	**2,125.6**	**2,392,272**	**11,424.6**	**5.4**
NCR	**636.0**	**1,985,299**	**9,453.6**	**14.9**
Manila	38.3	347,173	1,654.8	43.2
Mandaluyong	26.0	61,096	268.9	11.0
Marikina	38.9	73,617	357.2	9.2
Pasig	13.0	104,242	471.1	36.2
Quezon	166.2	415,788	1,989.4	12.0
San Juan	10.4	25,694	124.2	11.9
Kalookan	55.8	215,122	1,023.2	18.3
Malabon	23.4	74,657	347.5	14.8
Navotas	2.6	49,471	228.0	87.7
Valenzuela	47.0	94,377	437.2	9.3
Las Piñas	41.5	82,618	413.1	10.0
Makati	29.9	100,922	484.2	16.2
Muntilupa	46.7	80,981	399.8	8.6
Parañaque	38.3	82,692	391.3	10.2
Pasay	13.9	86,253	408.6	29.4
Pateros	10.4	11,377	55.3	5.3
Taguig	33.7	79,219	381.4	11.3
Cavite	**185.7**	**138,388**	**659.1**	**3.5**
Bacoor	52.4	52,594	250.6	4.8
Cavite City	11.8	20,059	92.6	7.8
Imus	97.0	36,846	177.4	1.8
Kawit	13.4	11,701	57.0	4.3
Noveleta	5.4	5,725	27.3	5.0
Rosario	5.7	11,463	54.1	9.5

Rizal	1,303.8	268,585	1,312.5	1.0
Angono	26.0	12,561	59.4	2.3
Antipolo	306.1	71,475	345.5	1.1
Baras	23.4	3,998	20.1	0.9
Binangonan	72.7	28,129	140.7	1.9
Cainta	10.2	40,671	201.6	19.8
Cardona	31.2	7,206	35.5	1.1
Jala-Jala	49.3	3,871	19.9	0.4
Montalban (Rodriguez)	312.8	16,759	79.7	0.3
Morong	37.6	7,322	36.0	1.0
Pililla	73.9	7,555	37.1	0.5
San Mateo	64.9	19,652	99.2	1.5
Tanay	243.4	14,042	69.2	0.4
Taytay	33.7	30,419	144.7	4.3
Teresa	18.6	4,925	23.9	1.3

Source: JICA 1992/National Statistics Office (1997)

Table 2: Water Charges of Selected Water Districts (P/m^3)

Water District	Average Tariff	Minimum Charge (P/conn)	Consumption Bracket (m³) 11-20	21-30	31-40	41-50
Metro Manila						
MWSS[a]	6.43	29.40 (47.30)	3.57 (5.42)	4.36 (6.29)	5.46 (6.82)	6.30 (8.42)
MWSS[b]	8.78	29.50 (47.00)	3.60 (5.42)	6.85 (9.00)	6.85 (9.00)	9.00 (11.36)
East Zone	2.32	7.78 (19.60)	0.95 (2.15)	1.00 (2.20)	1.00 (2.20)	2.37 (3.71)
West Zone	4.96	16.69 (29.40)	2.03 (3.33)	3.87 (5.36)	3.87 (5.36)	5.09 (6.70)
Metro Cebu		90.65	10.00	11.76	32.26	32.26
Baguio City		120.00	13.50	15.00	17.00	17.00
Metro Iloilo		80.00	8.00	8.80	10.40	10.40
Metro Siquijor		99.00	4.70	16.30	18.40	18.40
General Santos		50.00	5.60	6.08	7.04	8.00
Davao City		50.00	5.25	6.80	9.00	15.00

Notes
a MWSS tariff schedule effective July 16, 1995 until July 30, 1996
b MWSS tariff schedule effective August 1, 1996 until July 30, 1997
c The figures in parentheses denote the composite price, i.e., including CERA (P1.30 prior to privatisation and P1.00 after), and an environmental fee of percentage of base price.

Private Firms and Public Water

Table 3: Domestic Water Price Structure in Selected Utilities in the ASEAN Region, 1995 (US$/m³)ᵃ

	Average Price[b]	Consumption Bracket (m³)					
		1-10	11-20	21-30	31-40	41-50	51-60
Metro Manila							
MWSS[c]	0.23	0.11	0.14	0.26	0.26	0.34	0.34
	(0.31)[d]	(0.18)	(0.21)	(0.34)	(0.34)	(0.43)	(0.43)
East Zone	0.09	0.03	0.04	0.04	0.04	0.09	0.09
	(0.14)	(0.07)	(0.08)	(0.08)	(0.08)	(0.14)	(0.14)
West Zone	0.19	0.06	0.08	0.15	0.15	0.19	0.19
	(0.25)	0.11	(0.13)	(0.20)	(0.20)	(0.25)	(0.25)
Jakarta	0.61	0.16	0.16	0.16	0.31	0.31	0.35
Bangkok	0.31	0.16	0.16	0.16	0.22	0.23	0.25
Kuala Lumpur	0.34	0.17[e]	0.26	0.26	0.26	0.42	0.42
Singapore	0.55	0.39	0.39	0.56	0.56	0.82	0.82
	(0.62)	(0.46)	(0.46)	(0.63)	(0.63)	(0.89)	(0.89)

Notes

a Currency conversions are based on foreign exchange rates as of 1 July 1997, i.e., ₱26.384/ $1.00 .

b Refers to average price across all users.

c Effective August 1996 to July1997

d Figures in parentheses represent the composite price including a currency adjustment factor, and an environmental fee of 10 per cent of base price. For Singapore, figures in parentheses include sewerage charge.

e 0.17 applies to consumption up to 15 m3; 0.26 applies to consumption from 15 to 40 m3.

Source: McIntosh & Yñiguez (1997)

Table 4: Breakdown of Concession Fees West Zone (million pesos)

Year	Concession Fee 1[a]	Concession Fee 2[b]	Total Concession
1997	1,475	218	1,693
1998	2,047	445	2,492
1999	1,731	390	2,121
2000	1,424	378	1,802
2001	1,158	362	1,520
2002	1,067	454	1,521
2003	1,038	398	1,436
2004	839	396	1,235
2005	799	394	1,193
2006	688	392	1,080
2007	584	391	975
2008	252	389	914
2009	493	388	881
2010	425	387	812
2011	431	386	817
2012	444	385	829
2013	368	385	753
2014	343	426	769
2015	142	307	449
2016	133	317	450
2017	131	69	200
2018	132	57	189
2019	135	58	193
2020	138	59	197
2021	6.3	0	6.3

Notes

a Includes: i) 90 per cent of the aggregate peso equivalent due under any MWSS loan which has been disbursed prior to the commencement date (including MWSS loans for existing projects and the UATP project) on the relevant payment dates; plus ii) 90 per cent of the aggregate peso equivalent due under any MWSS loan designated for the UATP project which has not been disbursed prior to the commencement date on the relevant payment date; plus iii) 90 per cent of the local component costs and cost overruns related to the UATP project

b Includes: iv) 100 per cent of the aggregate peso equivalent due under any MWSS loan designated for existing projects which have not been disbursed prior to the commencement date and have been either awarded to third party bidders or been elected by the concessionaire for continuation; plus v) 100 per cent of the local component costs and cost overruns related to existing projects

Source: MWSS (1997a)

Table 5: Breakdown of Concession Fees, East Zone (million pesos)

Year	Concession Fee 1[a]	Concession Fee 2[b]	Total Concession
1997	164	134	298
1998	227	219	446
1999	192	240	432
2000	158	215	373
2001	129	203	332
2002	118	301	419
2003	115	260	375
2004	93	257	350
2005	89	255	344
2006	76	217	293
2007	65	217	282
2008	58	216	274
2009	55	215	270
2010	47	215	262
2011	48	214	262
2012	49	214	263
2013	41	213	254
2014	38	236	274
2015	16	160	176
2016	15	158	173
2017	14	56	70
2018	15	57	72
2019	15	58	73
2020	15	59	74
2021	0.7	0	0.7

Notes

a Includes: i) 90 per cent of the aggregate peso equivalent due under any MWSS loan which has been disbursed prior to the commencement date (including MWSS loans for existing projects and the UATP project) on the relevant payment dates; plus ii) 90 per cent of the aggregate peso equivalent due under any MWSS loan designated for the UATP project which has not been disbursed prior to the commencement date on the relevant payment date; plus iii) 90 per cent of the local component costs and cost overruns related to the UATP project

b Includes: iv) 100 per cent of the aggregate peso equivalent due under any MWSS loan designated for existing projects which have not been disbursed prior to the commencement date and have been either awarded to third party bidders or been elected by the concessionaire for continuation; plus v) 100 per cent of the local component costs and cost overruns related to existing projects.

Source: MWSS (1997a)

Table 6: Water Tariff Structure of the MWSS Before and After Privatisation for Residential and Semi-business Dwellings (P/m^3)

Blocks	Residential			Semi-business		
	Before	After		Before	After	
		MWC	MWSI		MWC	MWSI
First 10 m³	29.50*	7.78	16.69	49.50*	13.06	28.01
Next 10 m³	3.60	0.95	2.03	6.05	1.59	3.42
Next 20 m³	6.85	1.81	3.47	7.45	1.97	4.21
Next 20 m³	9.00	2.37	5.09	9.45	2.49	5.32
Next 20 m³	10.50	2.77	5.94	11.00	2.9	6.22
Next 20 m³	11.00	2.90	6.22	11.50	3.03	6.5
Next 50 m³	11.50	3.03	6.50	12.00	3.16	6.79
Next 50 m³	12.00	3.16	6.79	12.50	3.29	7.07
Over 200 m³	12.50	3.30	7.07	13.00	3.43	7.35

Note
* Per connection, otherwise ₱ per cubic metre

Source: MWSS (1997b)

Table 7: Water Tariff Structure of the MWSS Before and After Privatisation for Business I and Business II Establishments (₱/m³)

Blocks	Business I			Business II		
	Before	After		Before	After	
		MWC	MWSI		MWC	MWSI
First 10 m³	134.00*	35.36	75.75	145.00*	37.24	82.05
Next 90 m³	13.45	3.54	7.61	14.60	3.75	8.26
Next 100 m³	13.50	3.56	7.63	14.70	3.77	8.31
Next 100 m³	13.55	3.57	7.66	14.80	3.80	8.37
Next 100 m³	13.60	3.59	7.69	14.90	3.83	8.43
Next 100 m³	13.65	3.60	7.72	15.00	3.85	8.45
Next 100 m³	13.70	3.61	7.75	15.10	3.98	8.54
Next 100 m³	13.75	3.63	7.78	15.20	4.01	8.60
Next 100 m³	13.80	3.64	7.80	15.30	4.03	8.65
Next 100 m³	13.85	3.65	7.83	15.40	4.06	8.71
Next 100 m³	13.90	3.66	7.86	15.50	4.09	8.77
Next 200 m³	13.95	3.68	7.89	15.60	4.11	8.82
Next 200 m³	14.00	3.69	7.72	15.70	4.14	8.88
Next 200 m³	14.05	3.70	7.95	15.80	4.16	8.63
Next 200 m³	14.10	3.72	7.97	15.90	4.19	8.99
Next 200 m³	14.15	3.73	8.00	16.00	4.22	9.05
Next 500 m³	14.20	3.75	8.03	16.10	4.25	9.10
Next 500 m³	14.25	3.76	8.06	16.20	4.27	9.16
Next 500 m³	14.30	3.77	8.09	16.30	4.30	9.22
Next 500 m³	14.35	3.79	8.11	16.40	4.32	9.27
Next 500 m³	14.40	3.80	8.14	16.50	4.35	9.33
Next 500 m³	14.45	3.81	8.17	16.60	4.38	9.39
Next 500 m³	14.50	3.82	8.20	16.70	4.40	9.44
Next 500 m³	14.55	3.83	8.23	16.80	4.43	9.50
Next 500 m³	14.60	3.84	8.26	16.90	4.45	9.56
Next 500 m³	14.65	3.85	8.28	17.00	4.47	9.61
Next 500 m³	14.70	3.87	8.31	17.10	4.51	9.67
Next 500 m³	14.75	3.89	8.34	17.20	4.53	9.73
Next 500 m³	14.80	3.90	8.37	17.30	4.56	9.78
Next 500 m³	14.85	3.91	8.40	17.40	4.59	9.84
Next 500 m³	14.90	3.93	8.43	17.50	4.62	9.90
Next 500 m³	14.95	3.94	8.45	17.60	4.64	9.95
Over 10000 m³	15.00	3.95	8.48	17.70	4.67	10.01

Note
*Per connection, otherwise ₱ per cubic metre.

Source: MWSS (1997b)

NOTES

1. The author acknowledges the excellent research assistance of Debbie Gundaya.
2. Although water rights exist for Angat Dam water and the Water Code allows for compensation in cases of water rights transfer, short- and long-term reallocations of Angat water were made by administrative fiat despite competing uses of its water. A recent ADB study (1996) estimated that during the dry season, the economic value of raw water for urban use, as imputed from the value of consumer surplus per unit of raw water transferred, ranges from 2 to 5.7 pesos (₱2 - ₱5.7) per cubic metre for a 10 per cent and 20 per cent shortage, respectively. On the other hand, the income foregone by farmers ranges from ₱ 1.6 to ₱2.9 per cubic metre, suggesting positive net benefits of reallocation during drought situations. In case of long-term permanent water transfer, the estimated economic value to urban users is less than ₱2 per cubic metre; but the income foregone by farmers ranges from ₱1.3 to ₱2.1 per cubic metre.
3. Typically, the Angat Dam provides gravity irrigation for rice cultivation to about 24,000 hectares in the wet season and 27,700 hectares in the dry season.
4. The Filipino partner must have experience in one or more infrastructure businesses such as water supply, communications, power, construction, or real estate, which generates at least ₱1 billion (US$33 million) in revenue annually or involves ₱2 billion (US$67 million) in equity. The foreign partner, in turn, should have experience in all of the following: water supply treatment and distribution; wastewater treatment and sewerage services; metering, leakage control, and customer service and billing; and design and construction management for system expansion. It should also have had two year's experience of supplying potable water and sewerage services to areas with a population of at least 2.5 million, one million connections, and 10,000 km of main pipes. The foreign partner must generate $30 million in annual revenues from water and sewerage services and have investments of $1 billion in equity. Both the local and foreign partners must each be a single company (not an association of companies, though more than one firm may be allowed through a special purpose subsidiary).
5. The UATP augments the Angat Reservoir by diverting water from the Umiray River which drains its water into the eastern slope of the Sierra Madre. An additional bulk water supply of 800 million litres per day (mld) is expected upon completion of the UATP in June of 1999.
6. Despite the initial strong opposition to privatisation by MWSS employees, the specific provisions of the Agreement relating to employee matters such as hiring policies, mandatory severance payments, non-diminution of benefits, and employee stock option plans were accepted, and eased potential labour problems during the transition period.
7. This is how coverage is defined in Schedule 2 which presents targets by municipalities. In the text of the Agreement, the exclusion refers to users who obtain water from a legal source other than the MWSS system. As a rule, large-scale groundwater users such as private waterworks, commercial, and industrial establishments are required to obtain permits from the NWRB. In practice, the majority of large wells are not registered and hence are, technically speaking, illegal. In this case the two definitions of exclusions may not always be consistent.
8. This involves building a new dam and reservoir in the Kaliwa River basin in Tanay, Rizal, east of Metro Manila which can provide a total additional water supply of 1,900 mld.
9. In general, concessionaires are charged 90 per cent of the amortization of all existing MWSS loans which have been disbursed prior to the commencement date; and the total amortization of the foreign and local loans, local component costs and cost overruns of the UATP and other existing projects that have not been disbursed at commencement date.

10. The CERA was first imposed in 1984 to increase MWSS revenues and pay the additional amortization cost on its foreign loans as a result of the sharp peso devaluation in 1983. The CERA provision in the Concession Agreement is somewhat of a misnomer; it is a simple surcharge which is not linked to changes in the foreign exchange rate.
11. Namely: Ayala/International Water (United Utilities and Bechtel), Benpres/Lyonnaise des Eaux, Aboitiz/Cie Générale des Eaux, and Metro Pacific/Anglian Water.
12. If CERA and the environmental fee of 10 per cent are included, the average composite price in the East Zone in $0.14/m^3$ and $0.25/m^3$.
13. The East Zone requested an increase of ₱2.06/m³, but was granted only four centavos per cubic metre, plus 21 centavos for adjustment to inflation. The West Zone asked for an increase of 74 centavos and was allowed to raise the water tariff by 31 centavos per cubic metre, plus 53 centavos for inflation.
14. These projections assumed a higher rate of *per capita* household water consumption to account for suppressed demand due to water rationing and unbilled water that is actually consumed. In the case of David et al. (1998), a higher population growth rate and base year industrial and commercial water consumption were assumed.

6. Private Sector Participation in Water Supply and Sanitation: Realising Social and Environmental Objectives in Mexico D.F.

Lilian Saade Hazin[1]

6.1 CHARACTERISTICS OF THE WATER SECTOR IN MEXICO

Water Availability

The provision of water and sanitation services is a real and growing problem in Mexico. The mean annual rainfall is around 777 mm, equivalent to a volume of 1,530 km^3, and an annual runoff of 410 km^3. With a national population of 95 million, the annual per capita water availability is approximately 5,000 m^3, twice the world average.

Despite the high rainfall, water distribution is difficult and costly since the geographical distribution of water does not correspond to the distribution of the population and its needs. Over 76 per cent of Mexicans live in the northern and upland part of the country where only 20 per cent of water resources are located. In addition, this region contains 70 per cent of the country's industry and 90 per cent of the country's irrigated land, and generates 77 per cent of the country's GDP. Water supply is also unevenly distributed seasonally since annual rainfall is mainly concentrated in the summer period. Droughts occur approximately once every ten years with the areas in the northern part of the country being most affected (CNA 1994).

A total of 459 aquifers have been identified in Mexico, from which 24 km^3 of water is extracted annually. Approximately one-third of these aquifers is severely over-exploited (SEMARNAP 1996). (See Table 6.1 for an overview of the state of Mexico's aquifers.) The reasons for this over-exploitation are discussed below.

Table 6.1: The State of Mexico's Aquifers

Zone	Number of aquifers	Recharge (km³/year)	Abstraction (km³/year)	Availability (km³/year)	No. of aquifers over-exploited by at least 20%
North West	149	5.10	5.01	0.09	20
North	86	4.87	5.00	-0.13	20
North East	61	1.65	1.45	0.20	27
Balsas	92	8.16	7.40	0.75	29
Valley of Mexico	26	1.96	3.08	-1.13	3
South East	45	40.80	1.99	38.82	1
National	459	62.32	23.93	38.60	80

Source: SEMARNAP (1996)

Water Supply Networks

While Mexico has invested significant federal funds in the construction of public infrastructure, numerous problems remain. The administrative and pricing policies in Mexico have not been effective in ensuring that the increasing demand for water is met. In particular, revenue from water supply has only contributed a small part of the total cost of financing the system. As a consequence, the water sector has been increasingly dependent on the federal budget for its continued development. Furthermore, there are significant cross-subsidies between sectors. Water subsidies have been provided for agriculture and municipal users, while industry has paid a higher contribution.

Approximately 40 per cent of water supplied to homes is lost through leakage (Government of Mexico 1991). This leakage is mainly due to the poor maintenance of the infrastructure and inadequate installations in homes. In addition, the average cost of providing water services to the fast-growing population is increasing. In the period 1976-1994, the annual average investment needed to provide an additional cubic metre of water was more than double in real terms relative to the period 1950-1975. Increased demand began to outstrip the government's financial capacity.

According to the National Water Commission (CNA), in 1998, 86 per cent of Mexico's population had access to potable water and 72 per cent had access to sewerage services. Only 64 per cent of the rural population had access to potable water and around 32 per cent to sewerage services (CNA 1999).[2] The sectoral goals for the year 2010 are very ambitious. One of the aims of the government is to provide an additional 25 million people with access to piped water (95 per cent coverage) and an additional 30 million people with access to sewerage services by the year 2010 (88 per cent coverage)[3]. It is estimated that the amount required to accomplish the goals established for the sector is almost 148 billion pesos (around US$15 billion). The highest proportion (62 per cent) is estimated to be for investment in sewerage and wastewater treatment (Barocio Ramírez 1999). Given the size of the investment needed, private sector participation is emerging as an appealing option in certain cities.

Wastewater Treatment

The wastewater treatment capacity in Mexico is very limited. While approximately 230 m^3 per second of wastewater was generated in the country in 1996, only 10 per cent of effluents were properly treated (SEMARNAP 1996). According to studies conducted by the CNA, of the 218 basins covering 78 per cent of Mexico, 89 per cent of the total pollution load is generated in only 15 (Jaime Paredes 1997). Around 60 per cent of wastewater discharges are generated in the Federal District and in seven out of the 31 Mexican states. Studies also show that the Valley of Mexico and the Lagunera region are among the areas with the heaviest groundwater pollution (CNA 1994).

The limited wastewater treatment capacity results in adverse health impacts. The annual costs of diarrhoeal diseases caused by water and soil pollution, as well as by lack of sanitation and food poisoning, have been estimated at US$3.6 billion, making water pollution one of the major environmental problems in the country (Margulis 1994). In light of some of these problems, water authorities are implementing new policies to try to increase the wastewater treatment level in the country by the year 2010. The Water Programme 1995-2000 envisages an increase in the treatment of wastewater from 17 m^3 per second in 1995 (while the design capacity was 43 m^3 per second) to 82 m^3 per second by the year 2000. While it will be difficult to meet this goal with the recent financial crisis, significant funds are being invested to increase the wastewater treatment capacity in Mexico.

Institutional Framework

In recent years important changes have taken place in the institutional framework related to water provision in Mexico. The CNA was established in 1989 to co-ordinate efforts and guarantee the rational use of water resources. It was created as an autonomous agency attached to the Ministry of Agriculture and Water Resources and is the sole federal authority dealing with water management (Aguilar Amilpa 1995). The CNA is now attached to the Ministry of Environment, Natural Resources and Fisheries, created in December 1994.

The framework in which the CNA operates was transformed when the National Water Law (*Ley de Aguas Nacionales*) was enacted in December 1992, replacing the Federal Water Law (*Ley Federal de Aguas*). Its regulations were amended in 1994. The new law provides a modern regulatory framework for water management. It is combined with other laws at the federal level that deal with ecological and health-related issues, as well as other regulations at the state level. The main focus of this law is on concessions, permitting greater private sector involvement in the water industry.

At the state level, there is no uniform legal framework for issues such as water fees, investment, the participation of the private sector, and service cut-offs due to non-payment. In order to find solutions to this problem, the federal government has analysed the characteristics of the current state water laws. Based on the state laws that were considered the most advanced, the government has developed a model state level water law (*Proyecto de Ley Estatal tipo*). The model is intended to provide general guidelines for strengthening the management capacity of water operators and achieving more efficient water service provision. The adoption of this model is a decision to be made by each state, according to its own characteristics and with the approval of the municipalities. The state water law model includes guidelines for the development of master plans, the appointment of directors general of water utilities, greater user participation, and factors to be considered when setting tariffs. The model also contains important discussions on the feasibility of creating independent regulatory bodies at the state level as well as on the explicit recognition of the different types of service provider, including private and mixed companies.

Efforts have been made to foster decentralisation. Current legislation states that the municipalities - with support from the states where necessary - are responsible for the provision of water supply and sewerage services. However, most municipalities have limited resources, lack technical capacity and do not have the necessary infrastructure to collect revenues for the services provided. Moreover, since municipal administrations change every three years, and there is no immediate possibility of re-election, they have no incentive for long-term water planning.

6.2   WATER SUPPLY AND SANITATION IN MEXICO CITY

The Mexico City Metropolitan Area (MCMA) comprises two political-administrative entities: the Federal District, and the peripheral urban municipalities of the State of Mexico. The total area of the Federal District is approximately 1,504 km^2 (National Research Council 1995). The metropolitan area extends to the East, North and West of the Federal District into 17 municipalities of the State of Mexico, having a total area of 2,269 km^2. Approximately one-fifth of the Mexican population lives in the MCMA, with 8.8 million people in the Federal District (DGCOH 1996) and slightly more in adjoining parts of the MCMA in the State of Mexico.

In total, the area served by the common water distribution and wastewater disposal systems equals 1,287 km^2. Management of water and wastewater services within the MCMA is shared by the Federal District and the State of Mexico, which are each responsible for providing water and sanitation services within their jurisdictional boundaries. The CNA is responsible for providing bulk water to the service areas (National Research Council 1995). Of the 62 m^3/second of water received by the MCMA, the Federal District consumes 56 per cent and the peripheral municipalities of the State of Mexico 44 per cent.

The daily per capita rate of water supplied in the past five years has been on average 360 litres. While this is less than in the period 1984-1990 when 400 litres were supplied, it is still high relative to other cities in Mexico and the world (see Figure 6.1). This is partly attributable to the high leakage rate (37 per cent), but also due to the fact that prior to the introduction of the new system, the tariff structure gave users no incentive to save water. The daily per capita rate of water supplied in the State of Mexico is approximately 230 litres. Authorities attribute the larger per capita water use in the Federal District to the fact that the Federal District is more developed, with more commercial and industrial activities taking place. In addition, there are many private wells in the State of Mexico that are not included in the estimates (National Research Council 1995).

Two-thirds of the water supply comes from the aquifer underlying the city and the rest comes from external sources, mainly from the Cutzamala system. The Cutzamala basin incurs high pumping costs since it is located at a considerable distance from Mexico City (some 130 km) and lies at a lower altitude (1,000m below the city).[4] Mexico City has been rapidly drawing down the aquifer under the city. Because of the nature of the soil, the depletion of the aquifer has caused the city centre to sink by several metres. It has been estimated that average subsidence has been 7.5 metres over the course of the last century, with some areas falling by as much as two metres in the last decade (Noll et al., forthcoming). Over-exploitation is due to the city's high population and the consumption patterns prevalent in the area.

Figure 6.1: Daily Water Consumed per capita*

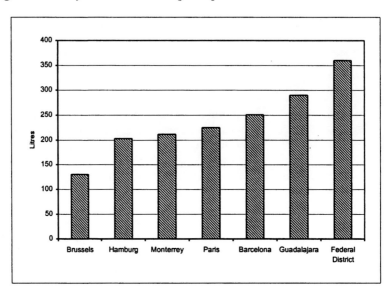

Note: * In the case of the Federal District water is supplied, not consumed

Source: Adapted from CADF (1994), p.4

According to the Federal District Water Commission, 98 per cent of the residents of the Federal District have access to piped water, either through an in-house connection or through a common distribution faucet in the neighbourhood. The remaining 2 per cent rely on water trucks from which they fill containers for home use. It is estimated that about 74 per cent of the residents of the Federal District have an in-house water source. In the metropolitan service area of the State of Mexico, approximately 52 per cent of homes rely on an in-house water source (National Research Council 1995). Figures from the CNA indicate that in the State of Mexico as a whole, 90 per cent of the population has some access to piped water. As shown in Figure 6.2, two-thirds of the water supplied to the Federal District is for domestic use, 17 per cent is used by industries, and the remaining 16 per cent is for commercial use.

Although 94 per cent of households in the Federal District has access to sewerage services, only 10 per cent of the municipal wastewater from the Mexico City Metropolitan Area (MCMA) is properly treated. There are 22 wastewater treatment plants in the Federal District and 16 in the State of Mexico service area, treating a total flow of 3,298 and 1,207 litres per second respectively. The latter represents around 50 per cent of their design capacity (CNA 1998b). The treated wastewater is used for local reuse projects such as groundwater recharge and agricultural and urban landscape irrigation (National Research Council 1995).

Figure 6.2: Distribution of Water Consumption in the Federal District

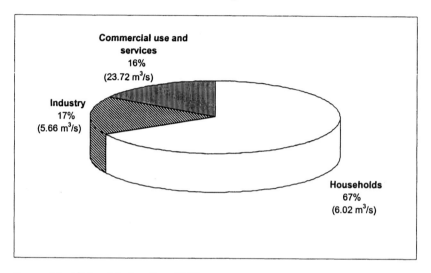

Source: Adapted from Martínez Baca (1996)

The remaining 90 per cent of wastewater generated is diverted out of the Basin of Mexico through the general drainage system. The untreated wastewater is then used to irrigate over 85,000 hectares of farmland in the Mezquital Valley in the neighbouring state of Hidalgo (National Research Council 1995). The main crops irrigated with wastewater in the area are fodder and cereal crops. A cross-sectional study of 2,049 households was conducted to test the impact of exposure to raw wastewater on diarrhoeal disease and parasitic infections in farm workers and their families. One of the main findings was that the risk of *Asvaris lumbricoides* infection is much higher in households where the farmworker is exposed to untreated wastewater than in the households where the farmworker practises rainfed agriculture (Cifuentes et al. 1993).

As in the rest of the country, the provision of water services in Mexico's Federal District has been highly subsidised. Since 1996, the service has required an annual subsidy of over 2 billion pesos ($US200 million) from other Federal District Government sources. The financial situation of water services is presented in Table 6.2. Expenses include investments in large projects that are also financed through external loans. With the recent changeover to metered consumption, more efficient physical and commercial operation and an increase in tariffs, it is anticipated that the situation will improve as revenues increase.

Table 6.2: Financial Situation ($USm 1998)[a] of Water Services in the Federal District

Year	Revenues[b]	Expenses[c]	Budget Deficit
1992	137	493	356
1993	158	424	266
1994	187	431	244
1995	143	387	244
1996	155	428	273
1997	183	518	335
1998	208	416	208

Notes

a The exchange rate at this time was 9.865 pesos/US$1.

b Revenues include water charges, wastewater discharge fees and water operation services

c Expenses include investments in large projects (sanitation for the Valley of Mexico, deep drainage, the Grand Canal (Gran Canal), water transmission line (acuaférico), water purifying plants, etc.

Source: based on information from CADF

6.3   PRIVATE SECTOR PARTICIPATION IN WATER SUPPLY AND SANITATION SERVICES IN THE FEDERAL DISTRICT

As noted above, the 1992 National Water Law encouraged private sector participation. This resulted in considerable private sector activity. In Aguascalientes and Cancún (both medium-sized cities), integrated concessions to private companies have been in operation since 1993. Service contracts were also signed in Navojoa in 1997 and in Puebla in 1998.

Of the 43 "Build-Operate-Transfer" (BOT) municipal wastewater treatment projects existing in 1997, only 12 have been built and are in operation, representing investments of $US170 million (Barocio 1999). Thirteen have been cancelled, and 18 are in the process of renegotiation. Of those in operation, most face financial problems and their treatment capacity is under-utilised. The main reasons for this are: the poor operating and financial performance of municipal water companies in general and their resulting lack of credit-worthiness, the inadequate structuring of contractual and concession arrangements, and the general procedures of the bidding process. Associated with these factors are gaps in the legal, regulatory and institutional framework at the state and municipal levels, as mentioned earlier. However, the primary concern of this chapter is the use of service contracts in Mexico's Federal District and it is this case which is examined in detail.

In July 1992, as a step towards solving the city's water problems, the Federal District Water Commission (CADF) was created. The CADF works as an autonomous body *(órgano administrativo desconcentrado)* under the supervision of the Works and Services Secretariat *(Secretaría de Obras y Servicios)* of the Federal District Government. The functions and responsibilities of the CADF will be discussed later.

In October 1992, the Federal District Government launched an international bidding process to select qualified private companies to conduct a census of users, install meters and rehabilitate the distribution system. The consortia were required by law to maintain majority Mexican ownership. In February 1993, seven firms submitted bids, of which four were selected by the Federal District Government. The commencement date of the operation was delayed because one of the unsuccessful consortia contested the awards, but the four successful consortia signed general contracts with the government in October 1993 (see Table 6.3).

The contracts were designed to overcome a number of deficiencies in the city's WSS system, including information gaps in the customer database, an inadequate tariff structure, and deteriorating water distribution and drainage infrastructures.

Each consortium was responsible for one zone (see Figure 6.3), with approximately the same number of water connections. In order to allow for certain economies of scale, zones containing adjacent municipal districts were awarded. It was considered that having four different companies would foster a certain degree of competition, and would ensure that the contractors used the best technologies available (Saade 1993).

Table 6.3: Consortia Involved in the Federal District's Water Distribution System

Zone	Company	Consortium partners[a]	Number of connections[b] (general intakes)	Contract value[c] (millions of pesos)
North West (Zone A)	Servicios de Agua Potable, S.A. de C.V.	Constructora ICA (Mexico); Vivendi (France)	350,064	969
North East (Zone B)	Industrias del Agua S.A. de C.V.	Socios Ambientales de México (Mexico); Azurix (USA)	290,000	981
South East (Zone C)	Tecnología y Servicios del Agua S.A. de C.V.	Peñoles (Mexico); Lyonnaise des Eaux (France)	304,059	979
South West (Zone D)	Agua de México S.A de C.V.	Grupo Gutsa (Mexico); United Utilities (UK)	291,990	880

Notes:
a The composition of the consortia has changed. In September 1999, Azurix acquired 49% of the capital stock of Industrias del Agua, S.A. de C.V. In January 2000, Peñoles acquired 51% of Tecsa's shares from Bufete Industrial.
b The number of connections represent general intakes as of the date the general contract was signed.
c Pesos are in constant figures as of March 1993, when the exchange rate was 3.11 pesos/US$1.

Source: Compiled from SOS (1998), Haarmeyer and Mody (1998) and personal communication with CADF staff

Figure 6.3: Allocation of Areas within the Federal District to Private Companies

Source: Adapted from CADF (1994), p.15

Given the lack of information regarding customer base, water consumption levels and the actual condition of the network, a phased approach to PSP was considered most appropriate. The specific tasks of the consortia were to be accomplished in three non-consecutive phases.

Phase I

During this phase, which started in May 1994, the companies installed meters, created a customer database and drew up network plans. The Federal District Government signed contracts with each of the consortia, in which the latter agreed to install a specified number of meters or carry out a census in a particular area. The Federal District Government paid the contractors per task accomplished on the basis of a unit price. By 31 December 1996, with the exception of meter installation, the first phase had been completed; 1.76 million users had been registered, and maps had been produced covering an urban area of approximately 680 km^2 (SOS 1998).

Two types of meter are being installed: conventional and electronic. By December 1998, more than 1.1 million meters had been installed. The total potential number of meters to be installed is around 1.4 million, although it will not be possible to achieve 100 per cent coverage since certain areas of the city have problems with water quality and continuity of service and will therefore require attention prior to the installation of meters. More than 20 per cent of consumers who still do not have meters pay a flat fee for water, regardless of consumption. However, this is a significant improvement on the situation prior to Phase I, when 50 per cent of households did not have meters.

Phase II

Phase II followed a contractual arrangement similar to that of the first phase, i.e., the Federal District Government paid the consortia for each task accomplished. Specific contracts under Phase II include meter reading; meter maintenance; assessment and distribution of bills; and new connections. By mid-1997, the second phase had made steady progress. Prior to 31 December 1996, the CADF (acting through the four companies) had been processing 72.4 per cent of all water bills, issued every two months, with the Treasury Department handling the remainder. By September 1997, the entire billing function for water services had been transferred from the Treasury to the CADF, with over 1.6 million bills being issued every two months. Table 6.4 shows the number of bills issued by the four companies by April 1998.

The delay in the implementation of these two phases was partly due to the economic and political situation that had prevailed in Mexico since December 1994, which affected the city treasury's financial situation. Many contracts depended on imported items, such as meters and other equipment. The effect of the 1994 devaluation was to more than double the price in pesos of these articles. In addition, there were budget cuts at all levels within government. Consequently, CADF had to delay or postpone several contracts.

In the case of Phase II, general contracts requested no specific rate to cover the cost of the activities related to customer care and bill collection. Tasks included the setting up and operation of a customer care centre for each 50,000 general intakes, a telephone information centre and field inspection teams. The contractors then considered the corresponding costs in the rates of other activities of Phase II for which a price was requested. Since no specific guidelines had been set for this cost allocation, the four consortia came up with different solutions. Some of them increased their rate for bill emission, others increased their rate for meter maintenance and one company distributed costs as a constant percentage of all its rates for Phase II.[5] This led to extreme differences between the contractors with respect to rates for Phase II for the same activities. Furthermore, the partial award of Phase II has caused important losses to the above mentioned contractor which distributed its customer care and collection costs as a constant percentage of all Phase II activities, only recovering 30 per cent of those monthly costs.

Table 6.4: Progress of Consortia through to April 1998

SAPSA	Azcapotzalco		Cuauhtemoc		Madero	Total
Bill Emission[a]	76,098		132,118		199,073	407,289
Installed Meters[b]	67,107		68,211		147,171	282,489
IASA	Benito Juarez		Coyoacan	Iztacalco	Carranza	Total
Bill Emission	103,392		147,830	72,775	86,807	410,804
Installed Meters	66,053		100,366	55,195	52,039	273,653
TECSA	Iztapalapa		Milpa Alta	Tlahuac	Zochimilco	Total
Bill Emission	321,444		12,717	50,539	57,529	442,229
Installed Meters	174,116		-	31,290	35,438	240,844
AMSA	Obregon	Cuajimalpa	Tlalpan	Contreras	Hidalgo	Total
Bill Emission	115,846	21,977	102,446	32,538	73,847	346,654
Installed Meters	93,095	13,829	71,487	14,453	59,803	252,677

Notes
a Billing: figures refer to bills issued in each two-month period.
b Meter installation figures indicate total installations.

Source: Personal communication CADF

Phase III

Phase III consists of the operation, maintenance and rehabilitation of the water distribution and drainage networks. It began, to a limited extent, in 1997. Its implementation was severely delayed as a result of internal political problems relating to the institutional framework (discussed below), and it was eventually split into two separate sub-stages. The first sub-stage already initiated with pilot projects has been temporarily suspended. It is based upon specific contracts between the CADF and the four firms. Firms will be responsible for sub-dividing their zones into District Meter Areas (DMAs). Flow meters will be installed at all entry and exit points of the DMAs, making it possible to calculate the quantity of water entering and leaving the area. Since the quantity of water consumed within a DMA can be established from readings of the micrometers installed during the first phase, the firms will be able to determine the quantity of water unaccounted for within the DMA. Unaccounted-for water is likely to be due mainly to physical leakage or illegal connections. By comparing the situation in various DMAs, companies should be able to determine where priority needs exist for extensive leak detection and, in some cases, rehabilitation of the network.

The second sub-stage of Phase III (which commenced in July 1997) comprises the operation and maintenance of the distribution network within the area of the DMAs that form part of the first sub-stage. Activities include the detection and repair of leaks, operation and maintenance of the water distribution network and its accessories (manhole covers and frames, valves etc.), and the replacement of valves and service pipes. While some of these activities were not included explicitly in the general contract, the contract had allowed for a certain degree of flexibility in order to allow for this type of activity to be carried out. By the end of 1998, the four companies had rehabilitated around 100 km of network, substituted more than 20,000 service connections (*ramales*) and identified and repaired leaks.[6]

It is still too early to evaluate the efficiency gains of PSP in Mexico City (see Table 6.5 for efficiency indicators in 1997). In order to be able to repay the contracted loans, technical efficiency (water distributed/water produced) will have to rise from the present level of 70 per cent. With respect to metering efficiency, this is expected to rise through the installation of meters and a collection system based on metered consumption in the city. Collection efficiency, which is currently around 65 per cent, is expected to increase through improvements in collection measures and customer service practices (Díaz 1997).

Table 6.5: Efficiency Indicators for 1997

Indicator	Value
Water system coverage	98%
Drainage coverage	94%
Daily per capita water supplied (litres)	360
Employees/1000 water connections	12
Employees/1000 users	8
Produced water (million cubic metres)	1104
Annual billing (million pesos*)	2377
Revenues from water supply fees (million pesos*)	1590
Efficiency (revenue/billed)	67%
Efficiency in metering (water metered/water distributed)	68%

Note: * The exchange rate at this time was approximately 8 pesos/US$1. Data in current pesos.

Source: Based on information from Haarmeyer and Mody (1998), and CADF

6.4   THE ROLE OF PUBLIC AGENCIES INVOLVED IN THE FEDERAL DISTRICT'S WATER SYSTEM

The framework for the public regulation of private sector involvement in the WSS sector was set out in the 1992 National Water Law. However, not all state laws considered the possibility of assigning concessions to private companies for the provision of water services. In order to overcome this problem the CNA's state level water law model is intended to provide general guidelines for the promotion of private sector participation.

In the case of the Federal District, the CADF was entrusted with the responsibility of engaging the private sector. It also maintains the database of users and is responsible for customer relations, establishing and maintaining plans for water distribution and drainage networks, and issuing water bills. Some of the CADF's most important functions are the issuing of bills, and the distribution of water to major users (*grandes usuarios*), who in 1992 totalled approximately 10,000. This number has now risen to more than 20,000. It should be noted that major consumers comprise only 1.4 per cent of all users in terms of numbers, but contribute more than 60 per cent of water fees collected.[7] The CADF's most important roles are those of supervising the four private consortia and reviewing and analysing their financial statements. However, a number of other public authorities continue to play significant roles in the sector:

The National Water Commission (CNA): is the main authority responsible for water management at the Federal level. It is an autonomous body (*organismo desconcentrado*) under the Ministry of Environment, Natural Resources and Fisheries (SEMARNAP). The CNA is responsible for providing water to the city.

The Federal District Government allocates functions to several public agencies:

The Directorate General for Waterworks Construction and Operation (DGCOH): is attached to the Works and Services Secretariat of the Federal District Government. It is the administrative unit responsible for receiving water and distributing it throughout the city. The DGCOH is also responsible for water purification and treatment plants throughout the city; supply wells and the arterial drainage system (including a network of deep tunnels that take surface and wastewater out of the city); and operation, maintenance and new works.

The Treasury Department of the Federal District: retains overall responsibility for the management of income generated from water, as well as the allocation of departmental budgets.

The 16 political municipal districts (delegaciones)[8]: are in practice in charge of the operation and maintenance of the secondary water distribution and drainage networks under the supervision of the DGCOH.

Finally, in accordance with the Presidential Decree of 21 October 1997 (Government of Mexico, 1997b) the **Works and Services Secretariat** (SOS), which is attached to the Federal District Government, is responsible for the operation and maintenance of the secondary water distribution and drainage networks. The Decree does not specify if this function will be undertaken by the CADF or the DGCOH but it is likely that the CADF will be given responsibility for this, with the work being undertaken by the four consortia.

The large number of public bodies involved in the regulation of the water operations causes certain problems. Although, in theory, each agency has a distinct role to play, in practice, there is a lack of co-ordination and there is often duplication and overlap of functions. In particular, the poor flow of information presents a potentially serious financial problem. At present, the four consortia are experiencing difficulty in calculating commercial efficiency in their respective areas (bills issued/bills paid) and pursuing debtors, since, although they know which customers have paid in their offices, they do not have ready access to complete information on payments made to banks or the various treasury offices.

To try to overcome some of the problems associated with lack of co-ordination, the CNA will continue to provide bulk water to the City but at the Federal District level it is intended to have only two agencies involved in water distribution in all stages - the DGCOH and the CADF (see Table 6.6). Both agencies are attached to the Works and Services Secretariat of the Federal District Government.

Table 6.6: Responsibility for the Water Distribution System in the Federal District

Activities	Current Resp.	Future Resp.
Transport of bulk water to the city and water purification	CNA	CNA
Dist. of bulk water to the primary network. Water distribution through underground wells. Operation of treatment plants	SOS through the DGCOH	SOS through the DGCOH
Water conn's greater than 15 mm in diameter*	SOS through the DGCOH	SOS through the DGCOH
Operation and maintenance of the secondary network (in transition)	Municipal districts	SOS most likely through the CADF
Updating the customers' register, installing meters and billing	SOS through the CADF (with the 4 firms)	SOS through the CADF
Billing. Revenue goes to the Treasury	Treasury	SOS through the CADF
Overall resp. for the sewerage system	SOS through the DGCOH	SOS through the DGCOH
Resp. for the deep sewerage system	SOS through the DGCOH	SOS through the DGCOH

Note: *The CADF is responsible for water connections of 15 mm or less.

Source: Personal communication from CADF

The role of the CADF as sectoral regulator is rather more straightforward for service contracts than it would be for more comprehensive forms of PSP. The companies' earnings have been derived solely from the completion of specific contracts with the CADF. The CADF also carries out physical surveys of the work undertaken by the companies. The General Contract contains a clause specifying that if any company performs poorly, either financially or technically, its contract can be revoked. There is the intention that the contracts will include a payment or penalty, depending on the improvement or deterioration in the level of physical leakage as calculated through the analysis of information obtained from the DMA. It will be interesting to see whether this is actually implemented and can be effectively regulated.

6.5   SOCIAL AND ENVIRONMENTAL ASPECTS OF PRIVATE SECTOR PARTICIPATION IN THE FEDERAL DISTRICT

Because of the nature of the service contracts, there are two primary social and environmental consequences arising from the involvement of the private sector in the Federal District's water supply system: conservation of scarce water resources, and greater access for poorer households. Through a household census of users, metering of connections, more effective billing, and rehabilitation of the distribution system it should be possible to manage water demand more effectively and ensure that poorer households do not face an excessive financial burden as costs of provision rise.

Water Conservation

In the case of the Federal District, the general tariff level for water is specified in the Federal District Financial Code. Tariffs are updated every year based on the recommendations of the CADF experts, the DGCOH, municipal districts, and the Treasury Department, and are approved by the Legislative Assembly (Government of Mexico 1997a). Prior to the implementation of the present water management scheme, tariffs in most places were fixed and highly subsidised (Casasús 1992). Despite some attempts through ecological information programmes to encourage Federal District inhabitants to use water more carefully, users considered water to be a cheap and unlimited resource.

In 1997 the Legislative Assembly approved the application of a new tariff system whereby every additional cubic metre consumed was charged for, rather than a flat rate being charged for consumption falling within a given range. However, in the Federal District's new water management scheme, responsibility for tariff setting remains with the Federal District authorities, although the private companies are often consulted. Moreover, the application of metered consumption to the majority of households has only been made possible through the census and metering tasks set out in the service contracts.

Following the establishment of billing and metered consumption, domestic water consumption in several zones has decreased by an average of 10 to 20 per cent.[9] However, it is important to note that even with these changes the current water tariffs still do not cover even 50 per cent of the cost of pumping water up into the city, let alone the scarcity value of water.

Wasteful water use was also exacerbated by the absence of a payment culture. However, there is some indication that this situation is changing. Although only approximately 65 per cent of all bills are paid, this is an increase of 30 per cent in the number of paying customers, due to the availability of more up-to-date information. According to the Federal District Financial Code, if domestic users fail to pay their bills, their water services can be rationed to the minimum level necessary to cover essential needs. In addition to problems of non-payment, the CADF estimated that only 63 per cent of the water supplied to the network was received by customers, some 37 per cent being lost through leakage.[10] The use of flow meters will help to identify leakages (and illegal connections). This, together with the extensive rehabilitation of the network, will reduce water losses.

However, significant problems remain. In particular, groundwaters continue to be exploited unsustainably, particularly in the eastern part of the city. This is mainly due to the fact that additional water sources from outside of the city flow in from the west. Since demand in the western and central parts of the city is met first, the external source is often exhausted before reaching the eastern part. As a consequence, four municipal districts in the eastern part of the city have to rely on local wells which are over-exploited and contain iron and manganese which give the water a poor appearance. The Federal District government has tried to alleviate this problem by distributing water to affected areas using road tankers.[11] However, this solution appeared inadequate and the Federal District government is now looking for a more permanent and socially acceptable solution through the construction of a water transmission line (*acuaférico*) that will transport water directly from west to east. Construction of the transmission line is expected to be completed by 2005.

Distributional Effects

As a basic need, demand for water tends to be income-inelastic. This is reflected in the ratio of financial expenditures (*gasto corriente monetario*) for water services per decile in urban Mexico. As can be seen in Figure 6.4, lower-income households spent a greater proportion of their total expenditure on water services. The chart is based on the National Households' Income-Expenditure Survey (*Encuesta Nacional de Ingresos y Gastos de los Hogares*) (INEGI 1994 and 1996). The last survey available relates to 1996. From Figure 6.4 it can be seen that in 1994 decile I (representing the poorest bracket) spent, on average, 4.5 per cent of its total expenditure on water services while deciles IX and X spent around 1.5 per cent. By 1996, the first decile spent almost 3 per cent of its total monetary expenditure while decile X spent only 1 per cent.

Figure 6.4: Percentage of Expenditure on Water Services of Total Expenditure by Decile

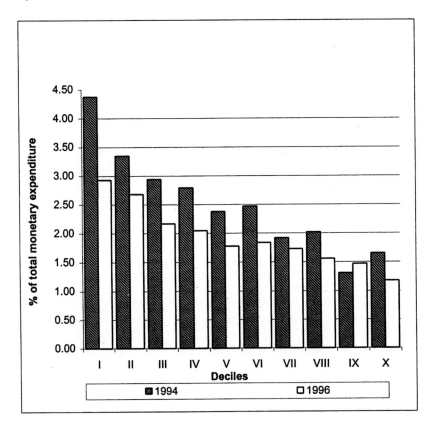

Source: Based on information from INEGI (1994) and INEGI (1996)

The service contracts may contribute to alleviating some of the financial burden for poorer households. Depending upon the tariff structure, metering and use-based fees may allow for the application of cross-subsidies from richer households to poorer households. While the rising block structure of the tariff schedule within the Federal District allows for this type of cross-subsidy, the real implications are not clear. Tariffs have been increased slightly (relatively) for residential users who consume less than 30m^3 in a two-month period and they have been increased considerably for users who consume more than 120 m^3 in a two-month period. For those who consume between 30 m^3 and 100 m^3 of water in a two-month period, there have not been significant changes in the tariffs they pay. In effect, between 1996 and 1998, they even enjoyed certain tariff reductions in nominal terms, which were applied in order to encourage customers to pay their bills. Residential customers consuming 30m^3 pay a *derecho* or charge of US$3.9, or US$0.13 per cubic metre (Government of Mexico 1999). Payments for water services are made every two months. There have been slight increases for users who consume between 100m^3 and 120m^3 (see Figure 6.5). In real terms, more than half of all residential users (using between 30m^3 and 220m^3 per two-month period) saw 1998 tariffs substantially lower than in 1996. This tariff reduction for residential consumers was accompanied by an increase in the cross-subsidy from non-residential to residential users (Haggarty et al. 1999). Since the consumption of non-residential users accounts for around 10 per cent of total consumption, the total effect has been that average real metered prices have decreased.

Such cross-subsidies only benefit households who are connected. According to the Mexican Human Rights Commission, the very poor currently pay five times more per unit of water than the average domestic water tariff, since they are less likely to have access to water in their homes and therefore have to rely on more expensive water vendors. Data on the provision of water services to the poorer households in the Federal District who do not receive piped water to their home are generally not available. Research conducted in some informal settlements showed that some residents were served by government tank trucks free of charge, other residents paid for an installed and metered network, and others paid a flat fee (National Research Council 1995). Rising block tariffs may be unfavourable to households that rely upon multiple-household connections, yet substantial efforts are being made to install meters per household and issue separate bills.

In the Federal District, retired citizens or residents over 60 years old who are able to prove their age can benefit from a 50 per cent discount. In addition, cancellation of charges for late payments and exemptions from debts incurred between 1995 and 1998 have been granted to residential customers who regularise their situation (Government of Mexico 1999).

Private Firms and Public Water

The CADF has faced criticism from certain groups about water services. In particular, several housing complexes where meters cannot be installed are unhappy with the new billing and collection systems. Others complain that the tariff increases have not been accompanied by significant improvements in service (SOS 1998).

Figure 6.5: Tariff Rates for Domestic Water Use in the Federal District

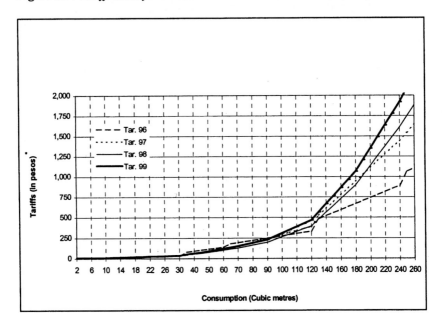

Note
* Data in current pesos

Source: Based on information from Federal District Financial Code for the corresponding years Government of Mexico (1996, 1997a, 1998 and 1999)

6.6 CONCLUSIONS

Private sector participation in the Federal District has been implemented since 1993 and a number of important lessons have been learned from private sector involvement in water in the Federal District:

- The results of local elections and political changes in the Federal District may have had an impact on the project. However, the need for this type of project in Mexico City clearly goes beyond political considerations. The new administration is aware of the importance of continuing the project. There is little question that metering, billing and rehabilitation have laid the groundwork for better management of scarce water resources.

- Performance under the contract has so far proved quite satisfactory. Despite several ambiguities, the contract anticipated and covered some changes, such as the transfer of responsibility for the secondary network and billing to the CADF.

- Important advantages were gained from the use of service contracts rather than concessions. A concession contract would have been far more complex and risky to implement than a service contract and would have required a more complex legal and regulatory framework. Despite the recent modifications to the legal and regulatory framework, there is still scope for regulatory improvement.

- The phased approach has allowed for the generation of information which is necessary for designing better contracts and for improving the regulatory capacity of the public authorities. The phased approach has allowed for sufficient flexibility for mistakes to be corrected and adjustments made to cater for unforeseen circumstances. A concession contract would not have allowed for the same degree of flexibility.

- Splitting the contract into four should have assisted the public authorities in regulating the sector since it allows for bench-marking of performance. In practice, however, the regulator has faced difficulties in comparing the performance of the firms. Since the project was conceived in three stages, some firms have charged more in the first phase, subsidising the second and third, and vice versa. Unit prices presented by the firms are not comparable to real costs. In fact, different municipal districts charge different prices in the same zone.

- Substantial changes to the institutional framework will be necessary in order to achieve more efficient water management. As the contractors expand the scope of their responsibilities and generate new information about consumers and the condition of their assets, further progress will depend, among other things, on the quality and response of the regulatory bodies.

- CADF's role of both contractor and regulator is a subject for further analysis and consideration should be given to the separation of these two functions.

- Finally, the potential benefits of PSP are also circumscribed by the scope of the contract. Given the strong interconnection, and the combined water and wastewater systems serving both the Federal District and the State of Mexico service areas in the MCMA, consideration has been given to creating a Metropolitan Water Commission to co-ordinate efforts. This is certainly a subject for further discussion, particularly since water is drawn from the same sources and deprivation is more pronounced in the peripheral states of the State of Mexico.

ANNEX 1: ACRONYMS

AMSA	Agua de México S.A de C.V.
CADF	Comisión de Aguas del Distrito Federal (Federal District Water Commission)
CAPA	Comisión de Agua Potable y Alcantarillado (Drinking Water and Sewerage Commission)
CNA	Comisión Nacional del Agua (National Water Commission)
DDF	Departamento del Distrito Federal (Federal District Department)
GDF	Gobierno del Distrito Federal (Federal District Government)
DGCOH	Dirección General de Construcción y Operación Hidráulica (General Directorate for Waterworks Construction and Operation)
IASA	Industrias del Agua
INEGI	Instituto Nacional de Estadística, Geografía e Informática (Statistics, Geography and Informatics National Institute)
MCMA	Mexico City Metropolitan Area
SAPSA	Servicios de Agua Potable
SARH	Secretaría de Agricultura y Recursos Hidráulicos (Ministry of Agriculture and Water Resources)
SEMARNAP	Secretaría de Medio Ambiente, Recursos Naturales y Pesca (Ministry of Environment, Natural Resources and Fisheries)
SOS	Secretaría de Obras y Servicios (Works and Services Secretariat)
TECSA	Tecnología y Servicios del Agua

ANNEX 2: CONCESSIONS IN CANCÚN AND AGUASCALIENTES

In addition to the service contracts in the Federal District, there have also been a large number of BOTs, and two concessions in Mexico. Since the latter is the most comprehensive form of PSP (almost a divestiture), this annex reviews the concessions in Cancún and Aguascalientes.

Cancún

The case of Cancún is interesting since in only 25 years the city has been transformed from a small fishing village on the country's Caribbean coastline into one of the most important tourist nodes in Mexico. Correspondingly, its population grew rapidly from 20,000 in 1976 to 340,000 in 1995 (Rivera 1996). This growth forced the authorities to look at new ways of increasing and improving the provision of services, including those of water and sanitation. In 1993, only 60 per cent of Cancún residents had access to piped water in their homes and 30 per cent had access to sewerage facilities. The 40 per cent of the City's population without household water connections had to rely on either public water taps or water trucks from which people filled containers for home use. Even among those who had access to piped water, some did not have a 24-hour supply. The 70 per cent of the population without access to sanitation facilities had to rely on the use of septic tanks and, in some areas, dumped waste in the open (Merino Juárez 1997).

Water and sanitation services were also unequally distributed between the hotel zone and the city, although both areas fell within the jurisdiction of the municipality of Benito Juárez. Priority was given to the hotel zone, while service provision in the rest of the city increased at a slower pace. Revenue derived from the hotel zone (which had a small, but lucrative client base) meant that, in contrast to many other water systems across the country, that of Cancún was profitable. Moreover, part of the revenue collected from the hotel zone was used to cover loans for previous investments and to subsidise other users. Despite the fact that qualitative and quantitative improvements were still needed in the Cancún area, some revenue was used to subsidise water systems in other municipalities.

Municipal and state authorities recognised the urgent need to improve the provision of water and sanitation in the area. Since their capacity to obtain loans was very limited, private sector participation was considered the most appropriate way of achieving this. In October 1993, the state government announced that municipalities had granted a concession to Desarrollos Hidráulicos de Cancún (DHC), S.A. de C.V. DHC, which would begin operating on 1 January 1994 through its subsidiary Aguakán. There was no public bidding process for the granting of the concession. The concession was signed for a 30-year period, with five-year targets and efficiency standards for the expansion and improvement of services. It covered the area incorporated in the Directive Plan for Urban Development in Cancún and Isla Mujeres. Clear definition of the area was important since new human settlements subsequently appeared which were not included in the Directive Plan (Merino Juárez 1997).

Under the concession agreement, the Drinking Water and Sewerage Commission (CAPA) was designated as the regulatory body responsible for revising tariffs and monitoring performance. However, the contract did not establish a system of penalties or fines for unsatisfactory performance or failure to comply with its conditions. The authorities did ultimately have the right to take away the concession but this could have had other significant implications and could only be done in exceptional circumstances. Nonetheless, in August 1996, the failure of the concessionaire to comply with its commitments, together with public criticism of the service, forced the government to temporarily take control of the private company. In February 1997, following a series of negotiations, the concession returned to the original concessionaire. In the interim period it was managed by a judicially appointed administrator.

In March of 1999 Azurix Corp, the water management subsidiary of Enron, acquired a 49.9 per cent interest from GMD in the water and wastewater concession for Cancún. This materialised after the modification of the concession agreement by the local water and wastewater regulatory agency subsequent to December 1998.

Aguascalientes

In Aguascalientes in central Mexico, a different set of problems relating to water supply are encountered, mainly concerning a low "natural" availability of water in the city. The growing deficit in the geo-hydrological balance constrains the state's development. In the past 30 years, the water level of the aquifer has decreased at an average rate of 2.7 m per year. Thus, the authorities have been forced to address the water shortage problem. They also opted for private sector participation.

As an initial step, a service contract was signed in 1989 with a consortium formed by Grupo ICA and Compagnie Générale des Eaux to provide water and sewerage services for a population of 750,000 (currently it is around 900,000). One of the characteristics of this contract was that revenues from collection would be given to the water operator. In May 1992, this aspect was modified and the private consortium became responsible for the administration of collected revenue. In 1993, the municipality of Aguascalientes (in the state of Aguascalientes) signed a 25-year concession contract with the consortium. As in the case of Cancún, there was no bidding process for the concession. This project has also been carried out in three stages: the first stage was a study of the city's situation; the second stage included marketing activities; and the third stage comprises operation and maintenance activities.

A number of factors, including the macroeconomic crisis of 1995 and a four-fold increase in interest rates in the same year, limited the concessionaire's financial capacity. The consortium responded by increasing tariffs by almost 60 per cent (Rivera 1996). The rate of inflation increased from 7 per cent in 1994 to 52 per cent in 1995. A number of political circumstances have also affected the implementation of the project. For example, in 1995 one of the opposition parties won the municipal elections. By the end of 1996, a new concession contract was signed, and the concession period was extended to 30 years. The service level has certainly improved. In the state of Aguascalientes, potable water is available to 95 per cent of the population and sewerage coverage to 89 per cent (Comisión Nacional del Agua 1999). Overall, significant improvements have been made in commercial efficiency. The contract was modified again in 1998 to adjust the investment programme and following the modification there were renegotiations over the investment agreements and questions were raised over the tariff formulas.

Among the lessons learned from this case are: the need for designing a "good" contract; the need for a regulator to sort out potential disputes; the need for local authorities to assert their political influence to effect tariffs (usually downward); and the intervention of the Federal Government to ease the fiscal strains of the concession.

NOTES

1. The author is employed in the Sector and Utility Management Group at IHE-Delft. She was formerly Advisor to the Deputy Minister of Planning at the Mexican Ministry of Environment, Natural Resources and Fisheries. The views expressed in this chapter are those of the author and do not necessarily reflect those of the institution to which she is attached. The author is grateful to Mr. Antonio Saade, Ms. Miryam Saade, Mr. Jean-Denis Hatt, Ms. Cioltzil Moreno, Dr. Nick Johnstone and Dr. Robert Hearne for their invaluable inputs. The author would also like to thank Mr. Jacques Couttelle, Mr. Michael Jones and Mr. Ramón Vila for their comments as well as Mr. Luis Díaz and Ms. Alicia Cárdenas for the information provided.
2. Rural areas are defined as those with less than 2,500 inhabitants.
3. The number of additional people to be provided with access to these services by the year 2010 is equivalent to the population of a country such as Morocco.
4. For a description of the water distribution system and the interconnection between the Federal District and the State of Mexico service areas in the MCMA refer to National Research Council (1995).
5. Personal communication with the contractors.
6. CADF, personal communication.
7. The four companies do not issue bills to major users.

8. The Federal District is subdivided into political and administrative units (*delegaciones*). At the end of 1997 the heads of the *delegaciones* were elected by the Legislative Assembly after being nominated by the Chief of Government and, in July 2000, they were elected by the capital's citizens by universal and direct vote.
9. According to estimates made by the contractors.
10. This figure applies to the Federal District. The estimated leakage level for the country as a whole is approximately 40 per cent.
11. In addition, efforts have been made to restrict urbanisation in the south-eastern part of the Federal District because of the difficulties in providing basic services, and also because it represents an important natural groundwater recharge zone.

7. Private Sector Participation in Water Supply and Sanitation: Realising Social and Environmental Objectives in Abidjan

Aké G.M. N'Gbo[1]

7.1 INTRODUCTION

Infrastructure services are of great importance to individual welfare and to economic development. Water, in particular, has a direct impact on health and as such access to water is considered a key human right. That right was reaffirmed at the International Conference on Water and the Environment held in Dublin in 1992, in the following statement: "...it is vital to first recognise the essential right of each human being to have access to potable water and hygiene at a reasonable price" (FAO 1997). Access to water is a particularly important issue in regions where water scarcity impacts on individuals and food production.

In less developed countries in general, and sub-Saharan African countries in particular, only a small proportion of the population has access to potable water (see Kerf and Smith 1996). Indeed, in 1991, only 44.9 per cent of the region's population had access to potable water, with considerable variability in access among countries. Recorded access rates varied from 7 per cent in Djibouti to 100 per cent in Mauritius. Rates for sub-Saharan Africa are lower than for East Asia and the Pacific (68 per cent) and for Latin America and the Caribbean (76 per cent). Compared to other sub-Saharan African countries, the population of Côte d'Ivoire is well served, with 83 per cent of the population having access to water (World Bank 1996). The low rate of access in sub-Saharan Africa is due to a number of factors, including insufficient network development, pricing problems, low availability of water, and competition between alternative uses. Water management is complicated by the mobile nature of water and the high cost of investment required to provide services.

Historically, water has been perceived as a good that could only be provided by the public sector. This is understandable given the high level of fixed costs associated with its supply and distribution. Moreover, there was concern that if service provision was left to the market, part of the population might be denied access to potable water. However, major constraints have arisen in the implementation of public water supply projects, including insufficient budgets and inefficient management on the part of the State. To overcome some of these difficulties, the private sector has become involved in water provision. In addition to having better access to capital, the private sector is also widely perceived as being able to manage the provision of these services more efficiently. PSP may take a number of forms, the most common being management contracts, leases and concessions.

The water sector in Côte d'Ivoire is somewhat different to other countries in that the private sector has been involved in water provision for a long time. This chapter analyses the water sector in Côte d'Ivoire. Section 7.2 outlines the institutional characteristics of the sector. In Section 7.3, the participation of the private sector is examined. In Sections 7.4 and 7.5 social and environmental objectives in light of PSP are discussed. The final section provides concluding comments and recommendations. Annex 1 provides an overview of the situation in rural Côte d'Ivoire.

7.2 INSTITUTIONAL, GEOGRAPHICAL AND PHYSICAL CHARACTERISTICS

Institutional and Statutory Aspects

The legal system in Côte d'Ivoire is strongly inspired by French law and traditional law and has its origins in pre- and post-independence legislation. In traditional law water was seen as a necessity, and as such, its usage, distribution and management were left to the community as a whole rather than to specific individuals.

Ownership of Water Resources

In Côte d'Ivoire almost all water resources are publicly-owned and therefore cannot be purchased. Publicly-owned water includes surface water, such as navigable water streams used for transportation (of timber, for example); non-navigable water streams, lakes, lagoons and ponds; navigation, irrigation and drainage channels; and aqueducts built for public use. It also includes water distribution systems, sewerage networks, harbours, rivers and the sea wall. Groundwater also belongs in the public domain.

However, there is some privately-owned water. For example, rain and stream water collected by individuals are considered to be private goods. Water from wells, drinking troughs and containers built by individuals on their land is also private, as is water from drainage and irrigation channels, and intermittent water streams. Moreover, by law, landlords have free usage rights. Residents who live adjacent to streams and are subject to the negative effects (on hygiene, aesthetic quality, public security etc.) from water distribution systems, communication and electric lines, and sewerage and drainage systems, do not receive any compensation.

Acquisition of the Usage Right

Private water can only be used by the owner of the land on which the water is located, unless prior authorisation, concession or permit has been obtained. The right to use private water can be acquired by donation, sale, inheritance or in light of any written contract relating to the land where the water is located. The use of public water, on the other hand, requires prior authorisation from the State.

As soon as an underground source is located, the Government acquires the land, which then becomes protected State property. In order to avoid deterioration of water quality, no human occupation and no activity whatsoever is permitted on or near these lands.

Water Use and Quality Regulation

The aim of regulation is to ensure the optimal usage of water resources and to protect the interests of those who rely upon water for competing uses (domestic, agricultural and industrial). Legislation pertaining to the regulation of water is quite embryonic. However, there are regulations covering the quality of water for domestic use. Regulation for other uses covers waste, quality, pollution and sanitation.

The Ministry of Housing and Living Environments' Inspectorate for Classified Installations (SIIC) implements pollution-related legislation, and is responsible for monitoring water pollution and air pollution emissions from industry and cars.

Protected Areas or Regions

Laws addressing protected areas or regions also exist. In the Abidjan region, a protected area has been established to ensure the conservation, preservation and sensible use of water resources. Exploitation of underground water within that area, or any work aimed at redirecting the water, is exclusively reserved for the public water distribution service.

Availability of Water Resources

Côte d'Ivoire does not face surface or groundwater availability problems (Sakho 1998 discusses water resources in Côte d'Ivoire). Total average rainfall for the country is estimated to be approximately 459 billion m^3 per year, of which 39 billion m^3 is available for use. Rainfall is characterised by wide spatial and temporal variability. Indeed, in an average year, rainfall varies between 2,400 mm in the extreme southwest and 950 mm in the northeast.

Groundwater reserves are estimated at 7 billion m^3 in the coastal sedimentary basin and at 78 billion m^3 in the crystalline platform. Usable resources are estimated at about 2.5 billion m^3 per year in the sedimentary basin and 3.5 billion m^3 per year in the platform area. Water streams are also found throughout the country.

Annual water consumption for all uses in Côte d'Ivoire is about 1.4 per cent of the total estimated capacity of around 1,080 billion m^3, with 100 billion m^3 allocated through the distribution network. Consumption for agricultural purposes is the highest, using about 950 billion m^3. Water consumption for different uses is summarised in Table 7.1.

Table 7.1: Annual Usage of Water

Usage	Billions of m^3	% of available resources	% of total consumption
Domestic	108	0.14	10
Industrial	25	0.03	2.3
Agricultural	950	1.2	87.7
Hydro-Electric	37 600		

Source: Sakho (1998)

Potable Water Distribution

The first potable water supply system in Côte d'Ivoire dates from the colonial period (Direction de l'Eau 1998a). However, until independence in 1960, only 18 communities had benefited from the development of formal water supply systems. In 1973, the National Programme for the Development of Potable Domestic Water Provision was implemented in urban and rural areas to address the problem of potable water supply.

Historical Background and Evolution

Until 1956, the water networks were managed directly through the municipal authority's technical services department. Since that time, management by delegation has been favoured, with the management of public water services being assigned to public enterprises such as EECI[2] or to private companies (see Zadi 1995, N'Gbo 1997 and Goueti Bi 1998a and 1998b for discussions).

In 1959, following an international auction, SAUR, a French private company, became Abidjan's water supplier. SAUR later became SODECI (an Ivorian-owned private company) and entered into management contracts with other municipalities. Until 1974, two enterprises, one public (EECI) and the other private (SODECI) shared responsibility for the water sector through different management contracts and concessions. SODECI has never been a state-owned enterprise.

During SODECI's initial period of involvement in the sector, major network improvements were carried out, especially in Abidjan. For example, pipes were installed in a more systematic manner, and treatment facilities and storage reservoirs were built. SODECI made substantial efforts to achieve economic profitability through a number of measures, including establishing a more systematic programme for installing meters, monitoring billing records, and developing a prompt consumer response service. Indeed, SODECI began recording profits from the beginning of its activities, and the number of service providers in Abidjan increased from 3,947 in 1960 to 29,907 in 1972.

In 1974, SODECI became responsible for providing water throughout the entire country and thus became the sole operator in the sector. The areas previously managed by EECI were assigned to SODECI from this period to 1987. From the point of view of competition, this made little difference since each company has been responsible for covering a specifically demarcated service area.

Finally, in 1987, a new 20-year concession contract was signed between the State and SODECI, granting SODECI monopoly rights over the provision of water. The existence of SODECI as sole operator in the water and sewerage sector was a choice made by the Ivorian authorities, informed by SODECI's efficacy. SODECI has continued to improve its performance over the years.

Actors in the Sector

The two principal actors in the water sector are the State and its agencies responsible for regulation of the sector, and the private sector, which is directly responsible for water supply.

SODECI is managed by Ivorians and has remained registered as a private company in Côte d'Ivoire. The capital structure of SODECI is as follows:

⇒ 53 per cent owned by local interests (with 3 per cent owned by the government, 5 per cent owned by the employees and 45 per cent owned by the Ivorian private sector)
⇒ 47 per cent owned by French interests (with 46 per cent owned by SAUR and 1 per cent owned by the French private sector).

SODECI's performance targets are set by the *Direction de L'Eau* (Water Board) and are based on the structure of the network and the needs to be met. SODECI's performance has been remarkable in terms of increasing coverage and improving management efficiency. The percentage of the population with access to potable water is estimated to be 87 per cent in Abidjan and 60 per cent in other urban centres. However, despite these relatively high rates, a proportion of the population remains without access to potable water. This is partly because the water network has not been extended to reach all communities and is also due to a lack of purchasing power on the part of some households, even in areas which are connected.

Private retailers supply water to those slum areas which are not connected to the network. However, the price of retailed water is significantly higher than the official price (private retailers are officially sanctioned by SODECI; this is described in more detail later).

7.3   PRIVATE SECTOR PARTICIPATION

As noted in the previous section, the private sector plays an important role in the production and distribution of potable water in Côte d'Ivoire through a concession contract with a single operator, SODECI, which is responsible for the supply of water to urban areas.

The 1974 Agreement

The 1974 Agreement stated that SODECI was responsible for the operation and maintenance of the State's water supply equipment at its own commercial risk, with its income coming directly from receipts.

The Water Board (*Direction de l'Eau*) and the Ministry of Public Works (*Ministère des Travaux Publics*) were responsible for elaborating and implementing the investment programme. SODECI had no obligation to invest in infrastructure. However, it had to maintain and operate any network expansion although it was not necessarily consulted about investments. SODECI was, however, guaranteed financial compensation in case actual consumption (and therefore revenue) was less than expected.

During this period, Abidjan was experiencing a rapid increase in population as a result of a period of economic prosperity and political stability. Rapid development presented a number of specific problems to SODECI in its obligations to supply water to the population of Abidjan. In particular, areas of haphazard housing were developing rapidly. These areas had few facilities and were often built illegally, without any title deeds to land. Running water was extremely rare. Although water vendors operated in these areas, shortages were common and water was relatively expensive. In many cases, residents of these areas were not eligible for service connections and could not, in any case, afford the price of water, which was billed on a quarterly basis. As a result, there were many illegal operations to divert drinking water (Lorrain 1997).

In order to overcome the difficulties experienced in poorer areas, SODECI adopted a number of alternative approaches. SODECI sought to exercise control over water vendors in poor areas through specific billing procedures while acknowledging their social utility. The company also supplied water for public standposts, installed by the State and located in very densely populated (and often illegally developed) urban areas. However, this often encouraged very high levels of water consumption and wastefulness, as well as maintenance problems, which led to heavy costs being borne by the local authorities.

In 1976, for social reasons, small-scale users were connected free of charge to the network. The connection cost was reimbursed to SODECI by the National Water Services Fund. This resulted in an increase in the number of low-income service subscribers, but at the same time placed substantial burdens on the provider, particularly in the form of high management costs relative to the quantity of water being consumed.

Between 1973 and 1980, SODECI underwent very rapid development and was transformed into a "model Ivory Coast company". The objective of this transformation was an organisational structure based on participative management with decentralised departments.

However, SODECI's operations suffered as a result of the economic crisis in Côte d'Ivoire between 1981 and 1987. Water consumption fell due to slower urban growth in relation to the network expansion being carried out, and due to the sharp decline in industrial activity which contributed to a drop in sales to the industrial sector. The financial situation gradually deteriorated. Investment decisions taken without consulting SODECI were based on very optimistic estimates, resulting in substantial state loans. When consumption was low compared to the level predicted, SODECI kept its financial guarantee. This situation caused a decline in investments and put an end to network expansion. A development tax that contributed to the Development Fund was added to water bills to subsidise small-scale users, since the National Water Services Fund was unable to absorb the full costs of small service connection installations. In 1987, a new agreement was signed in an attempt to overcome some of these difficulties.

The 1987 Concession

In 1987 a 20-year concession was awarded to SODECI. Probably the most important change with this new agreement was the improved co-ordination of investments and operating needs. Under the concession agreement SODECI became responsible for the development and submission of investment programmes to the Water Board for renewal, expansion works and social connections. SODECI's revenue guarantee was taken away to ensure that the difference between its part of the tariff and the price paid by consumers was allocated entirely to debt payment and investments.

An important aspect of the new concession agreement was that SODECI administered the investment funds itself under the contract. It can carry out all investments of a value less than or equal to 80 million Financial Community African Francs (FCFA), wholly financed by the investment funds. If the value is greater than FCFA 80 million, SODECI has the right to submit a proposal or organise an auction. Although similar, this new agreement is not strictly speaking a concession since SODECI does not have investment obligations.

The concession contract determines the rules, with SODECI's participation (as the private sector) being based on a number of fundamental principles:

- The State has overall control concerning the sector's general strategy and decides on specific issues, e.g. investments and pricing;

- The private sector provider must be managed efficiently and is obliged to achieve good results;

- Commitment to a social policy, with implementation of a water sale price realignment, free social connections, and progressive pricing.

The Institutional Structure

The following section describes the different actors involved in water provision and their roles. [3]

Regulation

Through its different technical ministries and specialised agencies, the State is responsible for formulating water policy, regulating the concession and financing some predetermined works.

The Ministry of Economic Infrastructure. Regulation of the sector is carried out by the Ministry of Economic Infrastructure through the Water High Commission. The Ministry is also responsible for defining water policy. It has executive authority over the sector and assumes ownership of the system. It decides on development and renewal works as well as prices and tariffs and is in charge of the development funds.

The Water Board. The Water Board is the technical department and has been delegated responsibility from the Ministry of Economic Infrastructure for, among other things, the technical and financial aspects of the concession for potable water distribution in urban areas, with technical assistance from the National Office for Technical and Development Studies (BNETD).

The Water Board is also responsible for monitoring water quality. Supported by the National Institute of Public Health (INSP) it ensures that World Health Organisation (WHO) guidelines are followed. The Board also checks that SODECI makes proper use of government installations. It decides on potable water supply projects on behalf of the Government and, in consultation with SODECI, determines the basic components of the water tariffs.

The Ministry of Economics and Finance is responsible for financial aspects of the sector and management of the sector's debt through the National Water Fund.

The Ministry of Housing and the Living Environment is responsible for urban sewerage disposal.

The Ministry of Health supports the Water Board with health education initiatives concerning water-related diseases.

The following ministries also play roles in the water sector

- The Ministry of Agriculture and Livestock Resources
- The Ministry of Defence
- The Ministry of Higher Education and Scientific Research
- The Ministry of Transport
- The Ministry of Energy
- The Ministry of Water, Forestry and the Environment

The High Commission for Water is responsible for the integrated management of water resources. However, in practice there is little co-ordination between these ministries (Direction de l'Eau 1998b). Improved efficiency in the sector requires clearer definition of the role and responsibility of each actor. Current thinking may soon lead to an integrated water resource management policy.

The operator

As operator, SODECI has responsibility for operation of the public service through the treatment, supply and distribution of potable water. It is also in charge of the maintenance of the network including all resulting costs. SODECI is responsible for management of subscriptions, connections, billing and repairs. It covers 75 per cent of the country (Direction de l'Eau 1998a).[4] Services provided by SODECI are paid for through a proportion of the amount paid by subscribers.

Finance and Tariffs

The financial principles of the sector

In order to execute the Domestic Water Programme, the water sector's financial policy is based on the following principles[5]:

- Financial equilibrium, in which receipts from subscribers must cover operating costs;
- The application of a uniform tariff;
- Operator's income based on operating and distribution costs;
- Taxes on tariffs for the National Water Fund and the Water Development Fund and on the concessionaire's income;

- Progressive pricing to reflect social priorities and discourage excessive consumption.

Sources of finance

There are special taxes which are collected for investment in the sector. The receipts are deposited in two funds:

- **The National Water Fund**, located at the "Caisse Autonome d'Amortissement" (CAA), is financed by an extra tax on water sales and is intended for the reimbursement of loans. As noted previously, this fund comes under the authority of the Ministry of Economy and Finance.

- **The Water Development Fund** (FDE) is financed by another tax on the sale price of water. It is administered by SODECI and is used to finance social connections, renewal works, network extensions and new investments. The FDE gives priority to financing welfare-oriented connections, which receive 50 per cent of its funds. This is followed by investment in upgrading (Table 7.2 provides a summary of investments in urban water carried out through the FDE from 1994 to 1998).

The FDE is administered by SODECI and is incorporated in SODECI's accounts. SODECI is also reimbursed for projects undertaken with its own funds. However, all withdrawals of money require the Water Board's approval. Although SODECI administers the FDE, it is not responsible for network extension. New works (eg, extension of the network) are financed through the Water Board and are subject to government approval. SODECI may, however, put forward network extension proposals.

If the financial resources of the FDE are inadequate, the Government intervenes through the Special Investment and Equipment Budget (BSIE) or through a foreign loan. If outside funding is not assured, financial difficulties on the part of the Government may result in the suspension of network expansion.

SODECI's revenues. These revenues are calculated on the basis of the operating costs of production and distribution with a contractual margin of 5 per cent for the operator. This creates a cost-plus type of pricing regime. However, the performance levels that SODECI must attain, combined with the business risk that it carries, discourage both under- and over-investment. Moreover, the Water Board, backed by BNETD, supervises investments and the execution of work.

Table 7.2: Investments in Urban Water Carried out through the FDE (repairs and new work in millions of FCFA)

Year	1994	1995	1996	1997	1998
Investment	2,200	1,600	1,200	3,040	3,200

Source: Direction de l'Eau (1998a)

The tariff structure. A uniform tariff is applied throughout the country. This results in high cost customers paying a price below the average cost of production in their area, while others pay a price higher than actual costs. Users in Abidjan, which has lower production costs, subsidise the other regions. The tariff has two parts: a fixed charge and a progressive charge.

Table 7.3 provides a breakdown of the tariff structure based on 1996 average prices (323.5 FCFA/m^3). The progressive tariff is set through five levels. For 1996, the fixed charge of FCFA 1,656 covered a consumption of 9 m^3.

Table 7.3: Tariff Structure in 1996 in FCFA

	SODECI's net revenues	Value added tax	SODECI's gross revenues	Dev'ment Fund	Additional tax - National Water Fund	User price
Level 1 0 to 18 m^3	144	16	160	10	14	184
Level 2 19 to 90 m^3	198	22	220	54	12	286
Level 3 91 to 300 m^3	198	22	220	197.5	46.5	464
Level 4 > 300 m^3	198	22	220	228	84	532
Consumption of the Administration	198	22	220	57	113	390

Source: SODECI (1997)

The Sector's Results

The results achieved by the operator are significant. The billing rate is 83 per cent, which is just below the performance target of 85 per cent. The recovery rate, i.e., payment from billed households, is 98 per cent, which is equal to the performance target.

According to the 1994 database of urban indices, in Abidjan 74 per cent of households lived in neighbourhoods supplied by a public potable water distribution system (BNETD 1994). Today, thanks to the vigorous connection policy pursued by SODECI, this has increased to 87 per cent of households. In the country as a whole, the number of districts benefiting from the network's water distribution has increased from 16 in 1960 to 529 in 1997. SODECI now covers 75 per cent of the country (Direction de l'Eau 1998a). However, most figures for those with access to potable water in urban African areas seldom distinguish between the proportion of consumers directly connected to the network, and those with indirect access (through retailers, neighbours etc.). A more accurate indicator of household access to potable water would be the number of households with their own private connection.

In an attempt to increase staff efficiency, SODECI has imposed productivity targets (including 300 meter readings a day per meter reader, and eight service connections per day per plumber), awarding bonuses to staff, promoting staff on the basis of merit, and establishing a corporate culture (including fringe benefits and emphasis on employee training programmes).

Table 7.4 demonstrates an improvement in the performance indicators in terms of centres equipped, length of the network, number of subscribers and production. These positive results are important in enabling the State to pursue the National Domestic Water Programme through:

- Recovering royalties to pay the sector's debt;
- New development using the sector's own finance;
- Access to potable water for a greater proportion of the population.

Social Issues in the Water Sector
Social issues relating to the water sector are extremely important. Since water is a basic need, the involvement of the private sector has raised concerns about part of the population not having access to water for financial reasons. As stated previously, responsibility for potable water supply in urban areas is assigned to SODECI. Social aspects concerning access to domestic water supply are dealt with through a number of measures.

Table 7.4: Performance Indicators

Indicators	Year					
	1974	1981	1993	1994	1995	1996
Centres in the network	38	139	409	409	410	411
Length of the network	922	6,000	10,000	10,300	10,800	10,900
Number of subscribers	40,071	130,478	275,000	293,000	318,000	345,000
Production (1,000 m³)	43,326	90,096	103,000	106,000	110,000	120,000
Volume billed	35,528	72,065	89,000	92,000	94,000	100,000
Billing ratio in %	82	80	86	84	83	83

Source: Goueti Bi (1998a), SODECI (1996)

Water Pricing
The retail water price depends on operating conditions. In principle this should lead to spatial differences in price. However, given that conditions for the supply of domestic water in Abidjan are highly favourable and that it accounts for over half of total supply, the uniform tariff contributes to the development of supply in areas outside Abidjan where the cost of supply is higher. In this respect, uniform pricing is an important aspect of social policy since, on average, other districts are poorer than Abidjan. In addition, while the "base rate" is uniform, the tariff structure has a social scale which allows lower income users to pay lower tariffs than others. Low volume users (<18m³) pay one-third as much as high volume users (>300 m³) (Noll, forthcoming).

SODECI has adopted a quarterly billing system, with the bill being settled in a single payment. This method of payment is one of the limitations of the welfare options offered to disadvantaged households. The infrequency of billing makes it difficult for households with low and irregular incomes to manage since they are unlikely to be in the position to save regularly.

Figure 7.1: Pricing Scheme and Performance Indicators

370,000 Subscribers

- Private   80%
- Public   20%

Pricing Structure

1996 Average price =   323.5 F/m³

Concessionaire's fee   182.6 F/ m³
Value added tax   20.35 F
Special water tax   120.55 F
Extra tax   44F
Water Development Fund tax   76.55 F

Operation and maintenance costs and investments

Performance targets
- Recovery rate   98%
- Billing rate   85%
- WHO standards
- 24-hour distribution

Investment
- Social connections
- Renewal works
- New centres
- Network extensions with State approval

- **Debt service**

- **Sanitation**

Source: Goueti Bi (1998a)

It is difficult to define an affordable price level in operational terms since an increase in the number of subscribers does not necessarily mean that tariffs are affordable. However, if one considers the high quality of the water service provided by SODECI and comparable tariff levels in neighbouring countries, it could be argued that tariffs in Côte d'Ivoire are relatively affordable (see N'Gbo 1997).

Connections and Charges
In order to improve accessibility to potable water, a policy was drawn up to provide incentives for connection to the public network. Thus, connections deemed "social" are free based on the following conditions being satisfied:

- The diameter of the water meter is no greater than 15 mm;

- The number of water points in the home is four or less (a water point being defined as an outlet where one may collect water, e.g. a tap);

- Connections are not used for commercial purposes.

Any household, regardless of where it is located, may benefit from subsidised connections if it fulfills these criteria.
   The policy has produced significant results. Indeed, as Table 7.5 demonstrates, the number of subsidised connections far exceeds paid connections, and is also increasing at a faster rate than paid connections.

Table 7.5: Increase in the Number of Connections over the Last Ten Years

No. of connections	Year						
	1987	1990	1992	1993	1994	1996	1997
Subsidised	14,681	19,468	15,381	12,689	20,246	25,094	30,334
Paid	1,117	2,140	2,900	3,112	2,020	1,261	1,500

Source: Direction de l'Eau (1998a)

Service Expansion and Provision of Alternative Services
Despite the efforts to improve access to potable water, a section of the population is still not directly connected to the domestic water network. In urban areas, households without access to the water supply network use other

sources, and vary their source of water supply according to its end use - reserving water from the pump or from the public fountain for drinking. Buying water on a day-to-day basis helps to reduce bulk spending, as do other forms of obtaining water such as collection of roof runoff. Water is bought and stored in metal or clay containers, thus facilitating control of the amount used and limiting wastage.   Some households rely on wells contaminated by ocean water, resulting in higher incidence of water-borne diseases (Noll et al., forthcoming).

There are two main types of collective connection for those who do not have household connections. First, a community may obtain a connection and employ a *fontainier* whose job is to sell water to community members. The receipts from water sales are used to settle operating bills. This type of connection is subsidised. Alternatively, such a connection can be obtained by a private individual who then functions as a retailer (known as a *yacoli*). However, in this case, the connection is not subsidised.

The second type of collective connection is where one household is a registered consumer and permits other households to have access to its water. Thus a household may be the registered consumer on behalf of a group of families. This household then receives a subsidy. This is the most common type of arrangement in rural areas with an interconnected water network and in urban *yard* housing, where a single water point is often installed. Alternatively, the registered consumer sells water illegally to neighbouring households. In both cases, registered consumers pay only administration costs and a deposit on consumption, amounting to a single payment of FCFA 18,750.

Finally, households may dig their own wells, purchase water from unauthorised vendors or rely on surface water. In 1994, households in Abidjan in neighbourhoods not covered by the water network (BNETD 1994) obtained water from the following sources:

- 68 per cent from public fountains or from private retail vendors who had extended the public distribution network;
- 23 per cent from wells;
- 6 per cent from village pumps;
- 3 per cent from surface water.

Table 7.6 provides a breakdown of the sources used by households in the Abidjan municipality of Port-Bouet.

Table 7.6: Principal Water Sources in Port-Bouet (Abidjan)

Water source	Percentage of population
Wells	10
Vendors	65
Wells + vendors	11
Public fountains	5
Private connection	8

As noted above, water retailers help meet the needs of many low-income consumers. They also help in the battle against unauthorised vendors and their existence generates employment. Water retailers pay a deposit to SODECI for an industrial usage connection. Retailers are located in peri-urban areas or slums and generally supply water to low-income consumers. However, a survey carried out in some districts in Abidjan shows that the average retail price per cubic metre is very high compared to SODECI's prices, as shown in Table 7.7. But the retailer allows the consumer to get the quantity of water needed without having to pay for the fixed costs. The information provided here was collected for the purposes of this study at various water retail points in Abidjan neighbourhoods. The prices given in this table are those of the informal market.

Table 7.7: Retail Sale Price

District	Average price per m³ (FCFA)
Yopougon	1,000
Anono	1,000
Abobo	1,000
Plateau maquis CCIA	2,000
Plateau maquis finance	1,300
Yopougon (water fountain)	430

The cost of water sold by official vendors also tends to be higher than for households with connections. This higher price is largely a result of the fact that Côte d'Ivoire has adopted a policy of graduated tariff brackets, prejudicing users relying on connections with large volumes. Bracket number four is known as the "industrial" bracket because users with this level of consumption are usually industries and retail vendors. SODECI carries out routine checks on retailers' meters to check they have not been tampered with.

There are also *bornes fontaines* in some districts. These consist of a 23-litre reservoir, a pumping system for refilling the reservoir with water and a system for dispensing water once the reservoir is full. *Borne fontaines* have one or two outlets made of PVC pipe. They are installed by the municipal authorities and are not the responsibility of SODECI. The main problem with *borne fontaines* is maintenance and refurbishment. Many of them do not work due to lack of spare parts. The price of water from *bornes fontaines* is fixed at FCFA 10 per 23 litres.

7.5   SANITATION AND ENVIRONMENTAL POLICY

Issues related to water use, provision of sanitation services and environmental quality fall under the authority of three different ministries. These ministries work independently without effective coordination between themselves, which makes it difficult to integrate sanitation and environmental aspects into the water sector.

Sanitation

In 1958, following a cholera outbreak, the Government sought assistance for the establishment of a sanitation plan for Abidjan. In 1976, the Government signed a contract with SODECI. This contract was renewed in 1988 and states that the Government must pay royalties to SODECI for maintaining sanitation facilities. A concession contract is currently in preparation under which users will pay SODECI directly for its services. In terms of payment for sanitation services, the water price will vary based on whether or not the user benefits from the sanitation infrastructure.

The installation of sanitation infrastructure is carried out by public authorities. There are currently four sewage treatment plants (two are secondary plants) and over 50 discharge centres in Abidjan. In other urban centres, the emphasis is on drainage. However, real estate companies have also constructed a limited number of drainage sites for houses built in the new district development programme. Some individuals also build private sanitation facilities.

Abidjan's sewerage and drainage system includes 756 km of piping and 156 km of single sewage collection network for the elimination of waste water, 524 km of underground piping and 472 km of open concrete pits and storm drains (SODECI 1997). In Côte d'Ivoire as a whole, 7.5 per cent and

30 per cent of the population have access to sewerage systems and sanitation facilities of acceptable standards, respectively (SODECI 1998) .

Socio-economic surveys carried out in 1992 have provided a general overview of the types of sewage disposal systems used by households in both rich and underprivileged areas (Table 7.8). At this time, the proportion of households in Abidjan connected to the public network was low (29.1 per cent). A high percentage of households used individual sewerage disposal (60.1 per cent), whilst 4.8 per cent of the population continued to use open spaces. Connection levels to the public network were particularly low in the municipalities of Treichville (1.9 per cent), Attécoubé (0 per cent), Abobo (14.3 per cent), Adjamé (23.0 per cent) and Port-Bouêt (26.2 per cent).

In Port Bouêt, which lies on the coast, 49 per cent of households are not connected to the public network and do not have their own sewerage systems. These households are predominantly located in vulnerable neighbourhoods and use the beach and sea for liquid waste disposal (wastewater and faecal matter). This practice has caused these areas to become polluted. The installation of sewerage systems is complicated by the fact that this neighbourhood is at the same level as the water table. In the north of the city, seepage from latrines is threatening the aquifer that is the city's main source of water (Noll et al., forthcoming).

Table 7.8: Types of Sewage Disposal used by the Population of Abidjan

Area	Public sewerage network (%)	Private disposal system (%)	Other (%)
Abobo	14.3	85.2	0.5
Adjamé	23.0	77.0	0
Attécoubé	0	98.2	1.8
Cocody	46.0	50.4	3.6
Koumassi	27.7	72.3	0.0
Marcory	42.9	57.1	0.0
Plateau	53.4	46.5	0.1
Port-Bouêt	26.2	24.7	49.1
Treichville	1.9	98.1	0
Yopougon	59.3	40.4	0.3
City of Abidjan	29.1	66.1	4.8

Source: Compiled from information provided by BNETD

As noted above, SODECI is actively involved with urban sewerage. It is responsible for the overall supervision of sewerage and drainage systems, and of clearing and maintenance works. Currently, 45 per cent of registered water consumers have access to the sewerage network, and 19 per cent have access to septic tanks and latrines. Thus, 36 per cent of registered consumers do not benefit from SODECI's sewerage.[6] Altogether, the sewerage facilities are capable of evacuating 37,000 m³ of wastewater, of which only 5 per cent is treated. 70 per cent to 80 per cent of this domestic and industrial effluent is discharged directly into the lagoon (which explains its current high level of pollution). Moreover, the sewerage system as a whole suffers from a lack of regular maintenance (i.e. drain clearance). Drains are constantly clogged with household refuse and bulky objects that obstruct the flow of effluent to outlets. Effluent thus overflows the networks and pollutes the natural environment.

Environmental Policy

Environmental policy is an important element within the water sector. The National Programme aims to provide good quality services to all users. Quality control is carried out in accordance with WHO standards at the production and distribution levels. At each level SODECI and the Water Board frequently carry out inspections. Samples of water are collected and tested in laboratories. However, there is a need for better co-ordination so that environmental aspects can be integrated into water policy.

The new Environment Law of 1996 covers a number of water-related issues including water pollution and the protection of water resources. The law specifies that the State's role in water management strategies includes preservation of water source quality, avoidance of waste, and increased access to water supply. It also states that wastewater management should no longer be carried out by the State, local communities or other bodies which are liable to produce effluents that damage the environment. This function could become the subject of a concession.

The regulatory arrangements specify the legal structure. However, the water policy is currently being reviewed through a national water policy document with the object of integrating all elements of the sector in order to achieve sustainable management of water resources. A water law is also being prepared which will incorporate the legal framework into national water policy.

7.6   CONCLUSION AND RECOMMENDATIONS

The case of SODECI in Côte D'Ivoire is an important example of a partnership between the state and private sector. It is instructive to note how a poor country with a relatively sparse population can, with the cooperation of the international private sector, provide reasonably good WSS services.

Analysis of the water sector in Côte d'Ivoire is summarised below:

- Since gaining independence 40 years ago domestic water distribution in Abidjan has been carried out by a private company. The level of private sector involvement has increased over time, culminating in the current Convention, signed in 1987, which is effectively a 20-year concession contract.

- Regulation of the sector is the responsibility of the State through the Water Board (Ministry of Economic Infrastructure) and the Ministry of Economy and Finance for the financial aspects. SODECI is the operator, responsible for the production and distribution of domestic water and is paid on the basis of the production costs and a margin.

- Financial policy for the sector is based on the principle of financial equilibrium, a uniform tariff structure and a graduated pricing system, which takes into consideration social needs and discourages waste.

- Benefits resulting from SODECI's participation are considerable in terms of network expansion, number of subscribers, billing ratio (83 per cent), recovery rate (98 per cent), and the national coverage rate (75 per cent). All of these indicators reflect improved access to potable water.

- Private sector involvement has not hindered the development of social policy aimed at disadvantaged people. Policies include adjustments to water prices, taking account of social strata and providing financial help with connections. There has been a steady increase in the number of subsidised connections.

- Private retailers provide water to households that do not have access to the network. The costs are relatively high but these retailers do satisfy the needs of this section of the population.

Despite generally positive results, some aspects of WSS provision could be improved. In particular:

- Cost-plus regulation can lead to difficulties. Indeed, the nature of the pricing regime does not offer incentives to reduce costs, and may result in misdirected investment. Moreover, this type of regulatory system requires considerable information about costs, which is very costly for the regulator.

- WSS in rural areas has not developed in the same way as it has in urban areas. It may therefore be beneficial to consider PSP in rural areas through the development of alternative technologies (water fountains, semi-water conveyance etc.) without necessarily extending the network, since this could be costly.

- SODECI should also improve its rate of recovery of administration bills so that it does not have to compensate for its loss by deducting part of what it owes to the National Water Fund.

- In the past, sanitation and the environment were considered separately, but with the increased participation of SODECI in sanitation and with the new agreement between the Government and SODECI, it is hoped that they will be better integrated within the national water policy and the water code.

- Extension and development of the network has not had adverse effects on water quality. However, until now, the environmental and water policies have been poorly co-ordinated. It is hoped that the water code in preparation will provide for an integrated framework for management of the water sector.

ANNEX 1: SERVICES IN RURAL AREAS

The National Village Water Supply Programme is an important element of the Water Supply Policy. This programme complements that of urban water supply assigned to SODECI. The African Water Supply and Renewable Energy Enterprise (SAHER), a privately-owned Ivorian company, is responsible for the provision of equipment and replacement materials for the water supply in rural areas. Village committees organise the management and maintenance of pumps.

Service Coverage

Responsibility for domestic water in rural areas is assigned to the Direction de L'Eau. The policy is to install potable water points in specific areas based on certain criteria:

- One water point for a population of 100 to 600 inhabitants;
- An additional water point for every additional 400 inhabitants;
- For populations of more than 2,000 inhabitants, an improved village hydraulic or semi-water conveyance is considered. This is an intermediate stage before network connection.

In areas that do not satisfy these criteria, the population uses wells or ponds. Water obtained from these latter sources does not generally meet WHO quality and hygiene standards.

A review of the National Village Water Supply Programme shows 70 per cent coverage of needs (Direction de l'Eau 1998a). According to the 1998 report of the Programme, there was a need for 19,871 water points in rural areas (Direction de l'Eau 1998c). As yet, only 15,637 have been installed and of these, 12,275 are in use. Financing has been assured for a further 2,929 water points. Thus, if one includes water points with guaranteed financing, the level of coverage will rise to 77 per cent.

Management of Water Points in Rural Areas

Village water points are managed and maintained by Village Management Committees (CVGs) which are responsible for meeting the villages' water needs. In addition, *fontainiers* are employed to distribute water and collect tickets at the various water fountains, their number being directly linked to the number of fountains. *Fontainiers* are not members of the CVG and receive a percentage of the takings from ticket sales.

Box 7.1:Village Water Management Committees

CVGs comprise six members: a co-ordinator and an assistant who are both government officials, a co-ordinator from the village, a treasurer, a technician and an assistant. The technician must have been educated to a certain level and must have some knowledge of plumbing and electricity. During construction they receive extra training from the companies responsible for carrying out the works. They are responsible for starting and stopping the system, reading water and electricity meters regularly, writing daily and monthly reports, and monitoring the network.

The village co-ordinator is in constant contact with the "Government" co-ordinator. In this way, the Government is kept informed of problems in the system. Along with the treasurer, the official co-ordinator oversees the sale of water-drawing tickets.

The treasurer is responsible for keeping and selling water-drawing tickets. The "Government" co-ordinator, selected from the officials closest to the village, is responsible for drawing up the monthly management report, discussing water matters with the people of the village and overseeing management. He reports to the government on the functioning of the system.

This form of community participation is one of the means being tried of developing infrastructure and services, particularly in rural areas. However, management problems tend to arise after the initial phase of the project and constitute the weak point of this type of participative venture. In addition, although the CVG receives training during the implementation stage of the water supply project, this training tends to be insufficient with the result that CVGs can find management of installations extremely difficult. Existence of these difficulties is reinforced by the fact that many rural people continue to get their water supply from rivers.

NOTES

1.   The author is a researcher at CREMIDE, University of Abidjan and was also PTCI Programme Coordinator. The author wishes to thank Mr Gogoua Amantchi and Amegan Tovi, both students of PTCI, for their assistance. PTCI is an inter-university graduate programme (combining management, economic research and teaching institutions in Africa). The author also wishes to thank Dr. N'Zue Felix who translated the final draft. Grateful thanks also go to SODECI and the Direction de l'Eau for agreeing to make information available.

2. The EECI is the Ivory Coast Electric Energy Company, a public enterprise in charge of power production and distribution during this time.
3. This section is based on Direction de l'Eau (1998a).
4. This level of coverage is taken to be the number of municipalities with piped potable water as a percentage of the total number of municipalities in the countries.
5. See Direction de l'Eau (1998a), N'Gbo (1997), Goueti (1998a and 1998b) for discussions.
6. Information provided by SODECI's Director of Sewage Disposal.

8.   Conclusions

Nick Johnstone and Libby Wood

8.1.   INTRODUCTION

PSP in the WSS sector has the potential to generate a number of social and environmental benefits for a number of reasons, including its potential to increase efficiency within the sector and increase much-needed levels of investment. Poorer households may gain access to affordable services from which they have long been excluded; adverse public health effects of inadequate service provision may be mitigated; wastewater collection and treatment levels may be increased; and, scarce surface waters and groundwaters may be conserved.

However, the extent to which these and other benefits are realised is dependent upon the type of PSP adopted, the means by which it is introduced and the effectiveness of the role played by public authorities and other organisations. While the case studies do not provide any hard and fast rules on how best to ensure that social and environmental objectives are realised, they do provide a solid basis upon which to ascertain which issues are most important and warrant further attention from public authorities, concessionaires, development agencies, NGOs and other bodies. These will be reviewed briefly.

8.2.   PRICING, SYSTEM REHABILITATION, AND RAW WATER SUPPLY

Despite the cost involved, a large number of contracts have included targets to increase metering of household connections. Along with the introduction of use-based tariffs (usually covering both water and sanitation), these measures provide incentives to conserve scarce water resources. Indeed in the case of Mexico City, where water shortages are particularly acute, the first two phases of the contract were largely designed to achieve this objective. Water consumption has fallen by as much as 10 to 20 per cent on average as a consequence. In relatively "water-rich" cities, such as Abidjan and Buenos Aires, few incentives are included in the contracts to increase metering and use-based pricing.

Most contracts also include targets for the rehabilitation of the distribution system in order to reduce water leakage rates. This is one of the primary objectives of the third phase of the contract in Mexico City. Both the Buenos Aires and Manila concessions include specific targets for reductions in "unaccounted-for water" (UFW). In Buenos Aires, UFW is targeted to drop from 45 per cent to 25 per cent by the end of the 30-year concession. In Manila, a reduction of UFW from 60 per cent to 30 per cent is required over a 25-year period.

While such targets are valuable, in many cases it may be more effective and less burdensome on the regulator to charge the service provider for the scarcity value of the raw water supply and let them determine what level of "leakage" is economically efficient. Despite the opportunity presented by the introduction of PSP this route is seldom taken. For instance, water from the Angat Dam near Manila is still allocated free of charge by government fiat, with the urban service providers given precedence in periods of scarcity. Imposing leakage reduction targets - while simultaneously providing free raw water to the concessionaire - is likely to be an economically inefficient and administratively costly means of conserving scarce water resources.[1] Moreover, the consequences for other users in the catchment area can be significant. For instance, following the drought associated with *El Niño* no water has been available for irrigation.

8.3   PROGRESSIVE TARIFFS AND NON-PAYMENT

In many cases, one of the keys to the realisation of social objectives in WSS is the application of a tariff schedule with positive distributional consequences. With a low income-elasticity of demand for water, this can be achieved through the use of rising block tariffs in which low levels of consumption of water are charged at lower rates. Thus, rising block tariffs are applied in Abidjan, Manila and Mexico City. In other cases (e.g. metered households in Buenos Aires and Córdoba), a lifeline tariff is used, where consumption up to a certain level is free. However, in these cases a monthly charge is applied in addition to the user tariff. This, along with the fact that free consumption is set at quite a high level, reduces the progressivity of the measures. In Buenos Aires, however, the level of free consumption is greater for poorer zones. For those households with both water and sanitation, user tariffs for sanitation are usually incorporated into the schedule.

The success of cross-subsidisation of metered consumption may be undermined by the tendency for water consumption amongst richer households with higher levels of consumption to be more price-elastic, since marginal uses tend to be less fundamental to household welfare. Moreover, where there are multiple-household dwellings with shared water points, rising block tariffs may even have perverse distributional consequences. This appears to be the case in Manila. Buenos Aires has tried to address this problem by adjusting the lifeline tariff for multiple-household dwellings.

Even in cases where consumption is not metered, other means are often used to cross-subsidise consumption for poorer households. For instance, Córdoba applies "zone" coefficients, in which households in poorer neighbourhoods pay slightly more than 50 per cent of the monthly charge of wealthier neighbourhoods. In Buenos Aires, a variety of proxies which are thought to be related to household wealth (construction date and "type" of dwelling) are also applied at the level of the household in order to determine monthly charges. However, there is likely to be a trade-off between the administrative costs and the effectiveness of progressive pricing in reducing inequities. Zone coefficients and other dwelling-related proxies are unlikely to be satisfactory guides to relative wealth, while "means-tested" measures are likely to be costly unless they can be "piggybacked" onto other social programmes with wide coverage.

The treatment of non-payment also has important social consequences. In many cities the threat of disconnection has often been used as one of the strongest incentives to increase payment collection rates. This has remained the case in many cities in which the private sector has become the provider. For instance, in Buenos Aires disconnection is possible for non-payment over three billing cycles. However, in other cases, public authorities have tried to restrict the use of this measure. Since private sector service providers are likely to have greater incentives to increase payment levels - indeed this is one of the perceived benefits of PSP - this may be necessary in order to ensure that poorer households with financial constraints do not experience excessive hardship. Determining the conditions under which disconnection is allowed for can involve trade-offs between efficiency objectives (increased collection rates) and equity objectives (ensuring that the poor are not adversely affected). Public authorities and concessionaires have tried to resolve this by various means. In Mexico's Federal District, in cases of non-payment the provider can restrict supply to the minimum level necessary to meet basic needs.

8.4   EXPANSION COSTS AND CONNECTION FEES

As we have seen in the case studies, due to historical biases against provision in poorer neighbourhoods, poorer households are often the greatest beneficiaries of service expansion and thus the social benefits of increased investment levels may be great. Moreover, expansion can also have environmental benefits for the wider community since households which lack access often adopt environmentally unsustainable practices for both water supply (e.g. excessive groundwater abstraction) and sanitation (e.g. seepage from latrines and unemptied septic tanks). However, the investment costs associated with service expansion can be considerable. Indeed, in many cases the costs will be even greater in poorer neighbourhoods since they may have been developed in an unplanned manner, located on the periphery, or be situated in areas with difficult topographical conditions.

The terms of many contracts have required private sector service providers to finance expansion costs from connection fees. This can be unaffordable for many poorer households with little or no access to savings or credit. One possible solution is to provide finance for credit schemes which effectively convert connection fees into monthly payments. In Buenos Aires, the concessionaire has to provide two-year financing for connection charges. Alternatively, it may be preferable to finance expansion costs from charges imposed on all users and not just new users. This is likely to be more equitable since in many cases users of the existing network did not pay for access when they were connected. The Buenos Aires case, where a surcharge on all users has been applied, is illustrative in this regard. In Abidjan, an investment fund is financed from a special water tax, effectively reducing connection charges.

Another solution could be to provide free "social" connections for poorer households. However, if the burden on the public purse is excessive this may have perverse distributional consequences, slowing expansion into poorer neighbourhoods. In the case of Abidjan, the number of free "social" connections far outstrips the number of new paid connections. One alternative which avoids this problem is the use of in-kind labour inputs as a substitute for financial payments in exchange for connection. Such a strategy has proved to be successful in Buenos Aires, actually accelerating service expansion rates.

8.5 FORMALISATION OF THE ROLE OF THE PROVIDER AND THE INCENTIVES OF THE PUBLIC AUTHORITIES

Perhaps the most significant benefit of PSP in terms of the realisation of social and environmental objectives is the formalisation of the role of the service provider relative to other agencies with an interest in these areas. The relationship between the service provider and the public authorities concerned with public health, housing, land use, water quality, wastewater treatment, groundwater abstraction and other related issues necessarily becomes more formal and explicit. This is important since trade-offs become more apparent, and priorities can be established in a less ad hoc manner (this point is made clear in the Argentinian case study). However, this is dependent on there being a degree of co-ordination between the different agencies that influence the regulatory environment of the sector. Responsibilities need to be established clearly at the outset.

PSP may also alter the incentives of the public authorities with responsibility for social and environmental objectives. Since their role becomes indirect, they may be more willing to introduce measures which increase costs of provision. For instance, public authorities with responsibility for water resources may be more likely to charge for raw water supply when it is a private sector service provider which is the buyer than when it is another public authority. They may also be more likely to introduce and enforce regulations on wastewater treatment and drinking water quality (it may not be a coincidence that new laws on water use and wastewater quality were introduced in Argentina soon after the concession was granted). If the measures are economically efficient and a reflection of social preferences then this is clearly beneficial. However, if they are not, this will not be the case. Indeed, when capital is scarce, there is likely to be a trade-off between the realisation of some environmental objectives (e.g. wastewater treatment) and other social objectives (e.g. service expansion), and the objectives of the various agencies have to be weighed carefully.

The separation of the objectives of the service provider and the consequent changes in the incentives of the public authorities also have the potential to increase accountability in respect of social and environmental objectives. While public service providers often have similar social and environmental objectives to those established contractually with PSP, public sector providers are less likely to be penalised when these objectives are not realised. However, it is important that the penalties provide appropriate incentives (i.e. are sufficiently high to affect firms' behaviour) but are also credible (i.e. are not so high as to discourage their application). For instance, some contracts only allow for the "dissolution" of the contract, which is hardly likely to be applied for minor infractions. Other contracts only allow for penalties which are of a lesser value than that of the infraction to the firm. Thus, the establishment of penalties appears to be a delicate balancing act. The use of performance bonds (as in Manila and Buenos Aires) helps to ensure that penalties are imposed.

8.6   FORMS OF PSP AND THE SCOPE OF THE CONTRACT

In most of the developing countries in which PSP is being introduced, conditions for the development of regulatory capacity are far from ideal. Rent-seeking, regulatory capture and technical constraints all pose problems. These can have adverse social and environmental implications since the authorities may not necessarily be in a position to ensure that related aspects of the contract are met. While every effort should be made to increase regulatory capacity, it is also important to design the contract in such a way that it eases the regulatory burden in relation to environmental and social objectives.

In the first instance, this may mean that it is necessary to adopt more limited forms of PSP which do not place such burdens on the regulator. Mexico City and Manila are at opposite ends in this regard. In the case of Mexico City, the phased approach to PSP through the introduction of service contracts and with the potential to gradually allow for increased management responsibilities over a 10-year period, gives the regulator the opportunity to progressively develop its capacity as PSP proceeds. Conversely, in the case of Manila, full concessions for water and sanitation were awarded at the outset. This places considerable responsibility on the regulatory authorities.

The spatial definition of the concession area can also be environmentally and socially significant. On the one hand, it will determine the potential scope for cross-subsidisation. In most cities (e.g. Buenos Aires and Abidjan) single tariff schedules are applied across the entire city, resulting in cross-subsidies from low-cost areas to high-cost areas. However, in Manila, the concession has been split in two, with tariff rates in the East Zone less than 50 per cent of those in the West Zone. Depending upon the relationship between the costs of service provision and the distribution of poorer households, this can have significant social implications. If concession areas do not cover a reasonable proportion of wealthier neighbourhoods it will be difficult to cross-subsidise poorer households.

8.7   CONTRACT DESIGN, REGULATORY BURDENS AND ILLEGAL CONNECTIONS

The relative weakness of regulatory capacity may mean that it is preferable to expand the scope of the contract in order to internalise incentives within the firm. For instance, by keeping sanitation in public hands in Córdoba it can be argued that an opportunity to internalise externalities directly through the contract itself was lost. With the sanitation service provider discharging wastewater into the water provider's raw water supply (Lago San Roque), a contract which covered both services might have been easier to manage since the firm would have appropriate internal incentives. Paradoxically, in the case of Buenos Aires where water and sanitation is not a "closed" system, the concession included both services.

In the case of Manila, the exclusion from the concession of users who rely upon private wells and other "legal" sources of supply is significant, particularly with surface waters in short supply and groundwater levels falling rapidly. Other "legal" sources of supply were allowed to continue following the granting of the contract to the concessionaires because it was believed that their efficiency would eventually allow them to undercut other providers and thus reduce excessive groundwater abstraction. However, this has not been borne out to the extent hoped. It may have been easier to ensure that water use was managed sustainably by expanding the contract to give the concessionaires full monopoly rights to service provision.

The means of introducing PSP should also account for the needs of households who rely upon illegal connections (direct or indirect). Since illegal connections have adverse implications for service quality (water leakage, pressure and quality) for the network as a whole, the incentives for the formalisation of illegal connections are clear. Since many poorer households are dependent upon such sources of water, it will not be equitable to close all illegal connections. Manila introduced an "amnesty" on illegal connections when PSP was introduced. Thus, efforts should be made to co-ordinate the removal or formalisation of these connections with service expansion into these areas. The same holds true for unregulated and unlicensed groundwater abstraction, which is a persistent problem in many of the case studies.

8.8   SERVICE PRIORITISATION AND DIFFERENTIATION

While most contracts have obligations for universal coverage, it is important to remember that in many cases full coverage is only achieved after a lengthy period. For instance, even if the terms of the contract are met in full in Buenos Aires, full water and sewerage coverage will not be attained for thirty years. Some neighbourhoods will not see the benefits of PSP until the concession itself is due for renewal. Since costs of provision are often higher and demand is lower in poorer neighbourhoods, it is likely to be the most disadvantaged who find themselves in such a position. This raises the question as to whether or not the obligations of the agreement should be prioritised by area, since poorer households may not be as well placed to adopt alternative measures which do not generate externalities.

It also raises the question as to whether or not "interim" measures should be mandated. For instance, the concession could include requirements for latrine and septic tank maintenance, as well as the provision of public water points or trucked water. The Manila concession includes requirements of this kind. The contract obliges the concessionaires to provide public standpipes and septic tanks in areas where household water connections and sewerage are not scheduled to be introduced in the near future.

While the introduction of interim measures is important, the ultimate objective in most contracts remains universal coverage of a standardised type of service provision. In some cities this may be appropriate but in many cases this objective may not be appropriate. Even with credit schemes, cross-subsidies and other financial support, poorer households may not be able and willing to pay for the same standardised services as richer households. By "vertically unbundling" the sector it may be possible to provide different types and levels of service in different areas. Thus, mandating technical specifications which provide households with services that do not reflect their actual preferences is neither efficient nor equitable.

This will, of course, be constrained by the need to ensure that household choices do not impose externalities on others, and by the technological constraints involved in the provision of collective services. However, in many cases some forms of service differentiation are clearly feasible and efficient. Thus, sub-contracting and sub-concessions should be allowed for with minimal transaction costs. In addition, the tariff structure should reflect different levels of service provision. Finally, some alternative forms of service differentiation may also help to overcome some of the problems associated with providing services in squatter areas. Since the contract for collective forms of service provision is not with the household itself, property rights are divorced from service provision.

8.9 USER PARTICIPATION AND DECENTRALISED MANAGEMENT

The issue of service differentiation raises important questions about the role of users in the design and implementation of PSP. Many contracts explicitly allow for rather simple forms of user participation. For instance, in the Argentinian case, complaints systems are formalised. This is significant since with split responsibilities it is important that households have a clear means through which they can express their dissatisfaction. Moreover, such systems will ease the regulatory burden on public authorities since households possess information on service quality which can only be obtained at considerable cost by others.

However, more fundamental forms of user participation may be advisable. For some stages of service provision (e.g. simplified sewerage or public water points), neighbourhood co-operatives and user associations may be vital to the efficient management of WSS services. By giving users a sense of "ownership" in investments the system is more likely to receive acceptance and therefore be efficiently and sustainably managed. The concessionaire in Buenos Aires has worked closely with users and NGOs in poorer neighbourhoods to try to make this possible. However, rather more needs to be done and there is an argument for formalisation of the arrangements. Since many NGOs and CBOs are better placed to build on existing social capital, they may have an important role in assisting poorer households to realise the benefits of PSP.

However, even prior to this stage, users should be involved in the design of the specifications of the contract itself. As noted above, this is a precondition for efficient development of the sector, particularly in poorer neighbourhoods, since it will ensure that households are willing and able to pay for the services provided. For instance, users can be provided with a set of service options (including cost and other implications). Few examples of PSP have allowed for such initiatives since it is perceived to be quite time-consuming and costly. However, in the case of Buenos Aires, consultation with users in the early stages would have saved both the service provider and the public authorities considerable effort, time and money.

8.10   BIDDING PROCEDURES, RENEGOTIATIONS AND RATE ADJUSTMENTS

Competitive bidding procedures are usually preferred when awarding contracts, since they are the primary means by which competition can be introduced in a sector characterised by significant economies of scale. Moreover, competitive bidding procedures increase the political legitimacy of awarding a "public" service to a private firm. However, there may be trade-offs between the transparency and legitimacy of the procedure and efforts to develop innovative means of solving persistent problems in the sector. For instance, in order for bids to be comparable, tenders often provide little leeway for proposing alternative technological solutions. Similarly, changing the specifications after the award has been made is often explicitly prohibited since this may undermine the legitimacy of the procedure.

Since there is likely to be little information but substantial uncertainty with respect to social and environmental factors, it may be necessary to forsake some of the attributes of competitive bidding to allow for projects which are flexible and innovative. This can be done through the use of "preferred bidders" in which contracts are awarded before the final specifications are established, through iterative development of the specifications in the bidding procedures. Another option - the route chosen in the case of Mexico City - is to adopt a phased approach, which allows for information to be generated as the project progresses and therefore provides more flexibility for eventual adjustments to be made.

Even with such procedures it will not be possible to foresee all eventualities, and it is in the environmental and social contexts that there is most uncertainty. Thus, there will need to be clear rules relating to adjustments once the contract has been awarded. For instance, in the case of Manila, there are at least four environment-related areas in which explicit provision is made for rate adjustments. These include: changes in service obligations (such as leakage rates or expansion targets); events of *force majeure* which increase costs (such as extreme droughts);[2] changes in laws and regulations (such as wastewater treatment levels); and, erroneous information provided prior to the bid (such as raw water quality - there was a dispute between the regulator and the concessionaire in Buenos Aires when it was discovered that the quality of water was worse than had been stated, and the cost of purification therefore increased).

8.11 CONCLUSION

Realising social and environmental objectives with increased PSP in WSS is, therefore, a complicated task. Table 8.1 provides an overview of some of the more direct means by which environmental and social objectives have been addressed in the case studies of private sector participation. However, as the case studies have shown, many of the most important effects can arise indirectly and incidentally. For this reason it is important that environmental and social objectives are fully integrated in the development of the strategy for private sector participation.

In general, all of the main actors (service providers, sectoral regulators, other public authorities, NGOs, and others) have a great deal to learn from each other. Moreover, with the speed at which PSP is progressing in the sector, the learning curve is very steep. Thus, every effort needs to be made to facilitate this learning. Firms need to gain a better understanding of social and environmental conditions in poorer neighbourhoods, public authorities need to get used to their role as regulators and not direct providers, and NGOs, the public and private sectors need to become accustomed to working together.

Table 8.1: Social and Environmental Market Failures and PSP-Related Measures

Source of market failure	Measures	Examples
Merit goods and preference failures	• Lifeline tariffs • Rising block tariffs • Property characteristics (unmetered) • Credit/financing schemes • Social connections and preferential tariffs • Service Differentiation	⇒ Buenos Aires, Córdoba ⇒ Abidjan, Mexico, Manila ⇒ Mexico, Buenos Aires, Córdoba ⇒ Buenos Aires ⇒ Abidjan and Córdoba ⇒ Manila and Buenos Aires
Raw water conservation and allocation	• Use-based water pricing (metered) • Metering targets • Leakage targets • Rehabilitation contracts • Withdrawal fees/permits	⇒ Manila, Córdoba, Abidjan, Buenos Aires[a], Mexico ⇒ Mexico ⇒ Manila, Buenos Aires, Mexico ⇒ Mexico ⇒ Mexico, Córdoba
Environmental and health externalities	• Quality targets • Treatment upgrading targets • Coverage/expansion targets • "Interim" Measures[b]	⇒ Buenos Aires, Abidjan ⇒ Buenos Aires, Manila ⇒ Buenos Aires, Manila ⇒ Manila, Abidjan

Notes
a In 1995 only 8 per cent of connections in Buenos Aires were metered.
b In the case of Buenos Aires a number of "interim" measures have been introduced but these were not explicitly included in the contract.

NOTES

1. However, in the case of Manila it may reflect a concern that increased water tariffs arising from pricing raw water supply may increase the tendency for households and firms to draw down scarce ground-waters even further.
2. Although the droughts caused by *El Niño* were considered a recurrent phenomenon, and thus foreseeable. As such, no rate adjustments were allowed.

References

Abdala, Manuel A. (1996), *Welfare Effects of Buenos Aires Water and Sewerage Services Privatization*, Washington DC: World Bank/Aguas Argentinas S.A./IIED América Latina.

ADB (1993), *Water Utilities Data Book - Asian and Pacific Region*, Manila: Asian Development Bank.

ADB (1996), *Water Management and Allocation Options: Angat River System*, Final Report of TA No. 2-PH, Colorado: prepared by R&D International.

Aguas Argentinas S.A./IIED-AL (1997a), *Optimización de la Programación y Expansión, Informe Final*, Convenio Aguas Argentinas/IIED-AL, Buenos Aires: Aguas Argentinas S.A./ IIED América Latina.

Aguas Argentinas S.A./IIED-AL (1997b), *Regularización de Barrios Carenciados en Areas Servidas de la Concesión. Plan de Acción 1997-98*, Convenio Aguas Argentinas/IIED-AL, Buenos Aires: Aguas Argentinas S.A./IIED América Latina.

Aguas Argentinas S.A./IIED-AL (1998), *Relevamiento Integral de Barrios Carenciados en el Area de la Concesión, Informe Final*, Convenio Aguas Argentinas/IIED-AL, Buenos Aires: Aguas Argentinas S.A./IIED América Latina.

Aguilar Amilpa, Enrique (1995), *Planning in Mexico: Experiences, Results and Perspectives*, paper presented at the Workshop on Intercomparison of National Water Master Plans March 27-29, Mexico: Comisión Nacional del Agua (CNA).

Altaf, M. J. (1994), 'The Economics of Household Responses to Inadequate Water Supplies', *Third World Planning Review*, 16 (1), 41-53.

AMEDC (Asociación Mexicana de Estudios para la Defensa del Consumidor, A.C) (1997), *Guía del Consumidor,* **27** (303).

Anderson, Dennis and William Cavendish (1993), *Efficiency and Substitution in Pollution Abatement,* World Bank Discussion Paper No. 186, Washington DC: World Bank.

Anton, Danilo (1993), *Thirsty Cities. Urban Environments and Water Supply in Latin America,* Ottawa: International Development Research Centre.

Argentine Government (1992a), *Decree 999/92, Concession Regulatory Framework,* Buenos Aires: Government of Argentina.

Argentine Government (1992b), *National Environmental Law Framework,* Buenos Aires: Government of Argentina.

Bahl, Roy W. and Johannes F. Linn (1992), *Urban Public Finance in Developing Countries,* Oxford: Oxford University Press.

Barocio Ramírez, Rubén (1999), *Participación del Sector Privado en el Sector del Agua Potable y Alcantarrillado en México,* paper presented at the Seminar on Tools of Private Participation in Water and Wastewater, Monterrey, Mexico, 10-12 March, 1999, Mexico: Comisión Nacional del Agua/Washington DC: World Bank.

Baron, D.P. (1985), 'Noncooperative Regulation of a Nonlocalized Externality', *Rand Journal of Economics,* **16** (4), 553-568.

Berlin, I. (1969), *Two Concepts of Liberty,* Oxford: Clarendon Press.

Bhattacharyya, Arunava, Elliot Parker and Kambiz Raffiee (1994), 'An Examination of the Effect of Ownership on the Relative Efficiency of Public and Private Water Utilities', *Land Economics,* **70** (2), 197-209.

Binnie/Thames Water/TGGI Engineer (1996), *MWSS Operation Strengthening Study,* Final Report of TA No. 2254-PHI, Manila: Asian Development Bank.

Bitran, Eduardo and Raul E. Saez (1994), 'Privatization and Regulation in Chile' in Barry P. Bosworth, Rudiger Dornbusch and Raul Laban (eds), *The Chilean Economy: Policy Lessons and Challenges,* Washington: Brookings.

BNETD (1994), *Computerised Cartographic Database of Abidjan Urban Indices, 1994.* Côte d'Ivoire: BNETD.

Boland, John J. and Dale Whittington (2000), 'The Political Economy of Water Tariff Design in Developing Countries: Increasing Block Tariffs Versus Uniform Price with Rebate', in Ariel Dinar (ed), *The Political Economy of Water Pricing Reforms,* New York: Oxford University Press.

Bourne, Peter G. (1984), *Water and Sanitation: Economic and Social Perspectives,* London: Academic Press.

Briscoe, John (1998), *The Financing of Hydropower, Irrigation and Water Supply Infrastructure in Developing Countries,* background paper prepared for the UNCSD, New York: United Nations Commission on Sustainable Development.

Brook-Cowen, P. (1997), 'The Private Sector in Water and Sanitation: How to Get Started' in *The Private Sector in Infrastructure: Strategy, Regulation, and Risk,* Washington, D.C: World Bank.

CADF (1994), *Agua: una Nueva Estrategia para el Distrito Federal.* México: Comisión de Aguas del Distrito Federal, p.4.

Cairncross, S. (1990), 'Water Supply and the Urban Poor' in Hardoy et al. (eds), *The Poor Die Young: Housing and Health in the Third World Cities,* London: Earthscan Publications.

Cairncross, S. and R. Feachem (1993), *Environmental Health Engineering in the Tropics,* Chichester: Wiley.

Cairncross, S. and J. Kinnear (1992), 'Elasticity of Demand for Water in Khartoum, Sudan' in *Social Science and Medicine,* **32** (2), 183-189.

Casasús, Carlos (1992), 'Una Nueva Estrategia de Agua para la Ciudad de México' in Ricardo Samaniego (ed.), *Ensayos sobre la Economía de la Ciudad de México,* Mexico: Librería y Editora Ciudad de México.

Casasús, Carlos (1994), *Private Participation in Water Utilities: the Case of Mexico's Federal District. Reinventing Government: New Opportunities for Public/Private Enterprise,* conference paper presented in Toronto, Canada, November 15 1994, Mexico: Federal District Water Commission.

CEPIS/ECLAC (1997), *Regulation of the Private Provision of Public Water-Related Services*, Peru: Centro Panamericano de Ingenería Sanitaria y Ciencias de Ambiente (CEPIS).

Chisari, Omar, Antonio Estache and Carlos Romero (1997), *Winners and Losers from Utility Privatization in Argentina,* World Bank Policy Research Working Paper 1824, Washington DC: World Bank.

Chogull, Charles L. and Marisa B. G. Chogull (1996), 'Towards Sustainable Infrastructure for Low-Income Countries' in C. Pugh (ed.) *Sustainability, the Environment and Urbanisation*, London: Earthscan.

Cifuentes, E., U. Blumenthal, G. Ruiz-Palacios, S. Bennett, M. Quigley, A. Peasey, and H. Romero-Alvarez (1993), 'Problemas de Salud Asociados al Riego Agrícola con Agua Residual en México', *Salud Pública de México* **35**, 614-619.

CNA (1993), *Situación del Subsector Agua Potable, Alcantarillado y Saneamiento a Diciembre de 1993*, Mexico: Comisión Nacional del Agua.

CNA (1994), *Informe 1989-1994*, Mexico: Comisión Nacional del Agua.

CNA (1997), *Estrategias del Sector Hidráulico*, Mexico: Comisión Nacional del Agua.

CNA (1998a), *Estrategia de Modernización del Subsector Agua Potable, Alcantarillado y Saneamiento*, Mexico: Comisión Nacional del Agua.

CNA (1998b), *Situación del Subsector Agua Potable, Alcantarillado y Saneamiento a Diciembre de 1996*, Mexico: Comisión Nacional del Agua.

CNA (1999), *Situación del Subsector Agua Potable, Alcantarillado y Saneamiento a Diciembre de 1998*, Mexico: Comisión Nacional del Agua.

Colby Saliba, B. and D. Bush (1987), *Water Markets in Theory and Practice*, Boulder, CO: Westview.

Cowan, Simon (1993), 'Regulation of Several Market Failures: The Water Industry in England and Wales', *Oxford Review of Economic Policy,* **9** (4), 14-23.

Crain, W. Mark and Asghar Zardkoohi (1978), 'A Test of the Property Rights Theory of the Firm: Water Utilities in the United States', *Journal of Law and Economics,* **21** (October), 395-408.

Crampes, Claude and Antonio Estache (1997), *Regulatory Tradeoffs in Designing Concession Contracts for Infrastructure Networks,* World Bank Policy Research Working Paper 1854, Washington DC: World Bank.

Crane, Randall (1994), 'Water Markets, Market Reform and the Urban Poor: Results from Jakarta, Indonesia', *World Development,* **22** (1), 71-83.

Dasgupta, Partha (1986), 'Positive Freedom, Markets and the Welfare State', *Oxford Review of Economic Policy,* **2** (2), 25-36.

David, Cristina C. (1997), *Water Demand Projections for Metro Manila: A Critical Review,* Policy Note No. 97-12, Makati City: Philippine Institute for Development Studies.

David, Cristina C. and A. B. Inocencio (1996), *Understanding Household Demand and Supply of Water: The Metro Manila Case,* Policy Note No.96-04, Makati City: Philippine Institute for Development Studies.

David, Cristina C. and A. B. Inocencio (1998), *Understanding Household Demand for Water: The Metro Manila Case,* research report, Singapore: Economy and Environment Program for South East Asia.

David, Cristina C. and A. B. Inocencio (1999), *Government Devolution and Privatization: Implications on the Water Service in Metro Manila,* unpublished paper, Makati City: Philippine Institute for Development Studies.

David, Cristina C., A. B. Inocencio, R. S. Clemente and G.Q. Tabios (1998), *Optimal Water Pricing in Metro Manila,* unpublished paper, Makati City: Philippine Institute for Development Studies.

Dellapenna, J. (1994), 'Why Are True Water Markets Rare Or Why Should Water be Treated as Public Property?' in M. Haddad and E. Feitelson (eds), *Joint Management of Shared Aquifers: The Second Workshop Nov.27-Dec. 1, 1994,* Jerusalem: The Harry S. Truman Research Institute/Palestine Consultancy Group.

DGCOH (1994), *Agua 2000. Estrategia para la Ciudad de México,* Mexico:

Dirección General de Construcción y Operación Hidráulica.

DGCOH (1996), *Análisis de la Situación de los Costos de Suministro de Agua Potable*, Mexico: Dirección General de Construcción y Operación Hidráulica.

Díaz, Luis, F. (1997), *Una Nueva Estrategia de Agua para el Distrito Federal*, paper presented at the Conference on Sanitation projects in Mexico, Mexico: Center for Business Intelligence.

Dillinger, William, (1994), *Decentralisation and its Implications for Urban Service Delivery*, World Bank Urban Management Programme Discussion Paper 16, Washington DC: The World Bank.

Direction de l'Eau MIE-RCI (1998a), *Alimentation en Eau Potable en Zone Urbaine - Bilan et Perspectives - 1956-2002*, working paper, Côte d'Ivoire: Direction de l'Eau.

Direction de l'Eau MIE-RCI (1998b), *Gestion des ressources en Eau - Evaluation des capacités nationales*, working paper, Côte d'Ivoire: Direction de l'Eau.

Direction de l'Eau, MIE-RCI (1998c), *Bilan du PNH villageoise*, working paper, Côte d'Ivoire: Direction de l'Eau.

Easter, K.W. and R. Hearne (1995), 'Water Markets and Decentralized Water Resources Management: International Problems and Opportunities', *Water Resources Bulletin* **31**, 9-20.

Electrowatt Engineering Services Ltd. and Renardet S.A. (1997), *A Project Review of Manila Water Supply III Volume I*, Philippines: Electrowatt Engineering Services Ltd. and Renardet S.A.

Estache, Antonio (1994), 'Making Public Infrastructure Entities Commercial', *Finance and Development*, September 1994, 22-25.

FAO (1997), 'Réformer les Politiques dans le Domaine des Ressources en Eau: Guide des Méthodes, Processus et Pratiques', *Bulletin d'irrigation et de drainage No. 52*, Rome: Food and Agriculture Organization of the United Nations.

Fass, S. M. (1993), 'Water and Poverty: Implications for Water Planning',

Water Resources Research, **29** (7), 1975-1981.

Feigenbaum, Susan and Ronald Teeples (1983), 'Public vs Private Water Delivery: A Hedonic Cost Approach', *Review of Economics and Statistics* **65** (4), 672-678.

Foster, Vivien (1996), *Policy Issues for the Water and Sanitation Sector*, IADB Infrastructure and Financial Markets WP 3/96, Washington DC: Inter-American Development Bank.

Franceys, Richard (1997), *PSP in the Water and Sanitation Sector*, DFID Occasional Paper No. 3, London: UK Department for International Development.

Gentry, Bradford and Lisa O. Fernandez (1997), *Evolving Public-Private Partnerships: General Themes and Urban Water Examples*, paper delivered at OECD Workshop on Globalization and the Environment, Nov. 13-14, 1997, Paris: OECD.

Goueti Bi, T. (1998a), *Case study of SODECI*, Seminar on public/private partnership in the management of urban water supply and sanitation in East Africa, Côte d'Ivoire: SODECI.

Goueti Bi, T. (1998b), SODECI presentation, working paper, Côte d'Ivoire: SODECI.

Haarmeyer, D. and A. Mody (1998), *Worldwide Water Privatisation: Managing risks in water and sanitation*, London: Financial Times Energy.

Haggarty, L., P. Brook and A. M. Zuluaga (1999), *Thirst for Reform? Private Sector Participation in the Urban Water Supply: The Case of Mexico City's Water Sector Service Contracts*, Mimeo, Draft 8, Washington DC: World Bank.

Haman, Bruno Z. (1996), *On Sustainability of Withdrawal from Metro Manila Groundwater System and Availability of Additional Groundwater Resources*, Makati City: Philippine Institute for Development Studies.

Hardoy, J., D. Satterthwaite and D. Mitlin (1992), *Environmental Problems in Third World Cities*, London: Earthscan.

Hearne R. and K.-W. Easter (1995), *Water Allocation and Water Markets:*

An Analysis of Gains-From-Trade in Chile, World Bank Technical Paper No. 315, Washington DC: World Bank.

Hearne, R. and J. Trava (1997), *Water Markets In Mexico: Opportunities and Constraints*, Environmental Economics Programme Discussion Paper 97-01, London: International Institute for Environment and Development.

Helm, Dieter (1986), 'The Economic Borders of the State', *Oxford Review of Economic Policy,* 2 (2), i- xxiv.

Helm, Dieter (1994), 'British Utility Regulation: Theory, Practice and Reform', *Oxford Review of Economic Policy*, 10 (3).

Helm, Dieter and Najma Rajah (1994), 'Water Regulation: The Periodic Review', *Fiscal Studies,* 15 (2), 74-94.

IDB (1996), *Argentina: El Proceso de Transformación del Sector de Agua Potable y Saneamiento,* Discussion Document, Washington DC: Inter-American Development Bank.

Idelovitch, E. and K. Ringskog (1995), *Private Sector Participation in Water Supply and Sanitation in Latin America,* Washington DC: World Bank.

Idelovitch, E. and K. Ringskog (1997), *Wastewater Treatment in Latin America,* Washington DC: World Bank.

INEGI (1994), *Encuesta Nacional de Ingresos y Gastos de los Hogares 1994,* Mexico: Instituto Nacional de Estadística, Geografía e Informática.

INEGI (1996), *Encuesta Nacional de Ingresos y Gastos de los Hogares 1996,* Mexico: Instituto Nacional de Estadística, Geografía e Informática.

Ingram, Gregory and Christine Kessides (1994), 'Infrastructure for Development', *Finance and Development*, September 1994:18-21.

Jaime Paredes, Alberto (1997), 'Water Management in Mexico: A Framework', *Water International: Official Journal of the International Water Resources Association, USA,* 22 (3).

Jaspersen, F. (1997), *Aguas Argentinas, The Private Sector & Development: Five Case Studies,* Washington DC: International Finance Corporation.

JICA (1992), *Groundwater Development in Metro Manila*, unpublished report prepared for the Metropolitan Waterworks and Sewerage System, Tokyo: Japan International Cooperation Agency.

Johnstone, Nick (1997), *Economic Inequality and the Urban Environment: The Case of Water and Sanitation*, Environmental Economics Programme Discussion Paper 97-03, London: International Institute for Environment and Development.

Johnstone, Nick (1998), *The Policy Implications of the Cost Structure of the Urban Water Supply and Sanitation Sector*, Briefing Paper prepared for DFID, London: UK Department for International Development.

Kerf, Michel and Warrick Smith (1996), *Privatizing Africa's Infrastructure: Promise and Challenge*, World Bank Technical Paper No. 337, Africa Region Series, Washington DC: World Bank.

Kerf, Michel, R. D. Gray, Timothy Irwin, Céline Lévesque and Robert R. Taylor (1998), *Concessions for Infrastructure: A Guide to their Design and Award*, World Bank Technical Paper No. 399, Finance, Private Sector, and Infrastructure Networks, Washington DC: World Bank.

Kessides, Christine (1997), *World Bank Experience with the Provision of Infrastructure Services for the Urban Poor*, Washington DC: World Bank.

Kinley, David (1992), *KWAHO's Urban Challenge*, New York: United Nations Development Programme.

Laffont, Jean-Jacques and Jean Tirole (1993), *A Theory of Incentives in Procurement and Regulation*, Cambridge, MA: MIT Press.

Lambert, David K. and Dino Dichev (1993), 'Ownership and Sources of Inefficiency in the Provision of Water Services', *Water Resources Research*, **29** (6).

Lewis, Maureen and Ted Miller (1987), 'Public-Private Partnership in Water Supply and Sanitation in Sub-Saharan Africa', *Health Policy and Planning*; **2** (1), 70-79, Oxford: Oxford University Press.

López Roldan, Raúl (1999), *La Perspectiva en México del Socio Privado: La Concesión en Aguascalientes y el Contrato de Servicios en el Distrito Federal*, paper presented at Seminar on Tools of Private Participation in

Water and Wastewater, Monterrey, Mexico, 10-12 March 1999, Washington DC/Mexico: World Bank and Comisión Nacional del Agua.

Lorrain, D. (1997), *Urban Water Management. French Experience around the World*, Paris: Hydrocom Editions.

Luther, David Scott (1993), *IDDI: Integral Development in the Dominican Republic. Voices from the City*, Arlington, Virginia: Environmental Health Project.

Lyonnaise des Eaux (1998), *Alternative Solutions for Water Supply and Sanitation in Sectors with Limited Financial Resources*, Nanterre: Lyonnaise des Eaux .

Mara, Duncan (1996), *Low-Cost Urban Sanitation*, Chichester: John Wiley.

Margulis, Sergio (1994), *Back-of-the-envelope Estimates of Environmental Damage Costs in Mexico*, revised version of World Bank Policy Research Working Paper No.824 written in January 1992, Washington DC: World Bank.

Martínez Baca, Alfonso (1996), *Analisis de Casos. La Participación Privada en Sistemas de Agua y Saneamiento: México Distrito Federal*, paper presented at the XXV Congreso Interamericano de Ingeniería Sanitaria y Ambiental, 2 November 1996, Mexico City: Comisión de Aguas del Distrito Federal, p.12.

McIntosh, Arthur C. and C. E. Yñiguez (eds) (1997), *Second Water Utilities Data Book: Asian and Pacific Region*, Manila: Asian Development Bank.

Merino Juárez, Gustavo (1997), *Mexico: Private Provision of Water in Cancun*, Cambridge, MA: Harvard University, Kennedy School of Government.

Mexico, Government of (1991), 'Programa Nacional de Aprovechamiento del Agua, 1991-94', *Diario Oficial de la Federación* 5 December, Mexico: Secretaría de Agricultura y Recursos Hidráulicos (SARH).

Mexico, Government of (1996), *Leyes y Códigos de México: Código Financiero del Distrito Federal*, Mexico: Editorial Porrúa.

Mexico, Government of (1997a), *Leyes y Códigos de México: Código*

Financiero del Distrito Federal, Mexico: Editorial Porrúa.

Mexico, Government of, (1997b), *Diario Oficial de la Federación* 21 October 1997, México.

Mexico, Government of (1998), *Leyes y Códigos de México: Código Financiero del Distrito Federal,* Mexico: Editorial Sista.

Mexico, Government of (1999), *Código Financiero del Distrito Federal,* Mexico: Editorial Sista.

Mody, Ashoka (1996), 'Infrastructure Delivery: New Ideas, Big Gains, No Panaceas' in A. Mody (ed.) *Infrastructure Delivery: Private Initiative and the Public Good,* Washington DC: Economic Development Institute, World Bank.

Morales-Reyes, Javier, Richard Franceys and Kevin Sanson (1998), *Mexican B.O.T. Experience and World Comparison Analysis and Recommendations.* WEFTEC '98, Workshop 109, Alexandria, VA: Water Environment Federation.

Moss, Jack (1997), 'A Contract for the Production of Drinking Water for Sydney' in D. Lorrain (ed.) *Urban Water Management: French Experience Around the World*, Paris: Hydrocom Editions.

MWSS (1997a), *Concession Agreement,* unpublished document, Manila: Metropolitan Waterworks and Sewerage System.

MWSS (1997b), *New Water Tariff Rates for the East and the West Zone,* Manila: Metropolitan Waterworks and Sewerage System.

N'Gbo, A.G.M. (1997), *Universal Service in Infrastructure Services: A Survey of the Côte d'Ivoire Experience,* paper presented at World Bank Seminar, Nov. 1997, Washington DC: World Bank.

National Hydraulic Research Center (1993), *Water Resources Management Model for Metro Manila,* Final Report to International Development Research Center, Manila: University of the Philippines.

National Research Council (1995), *Mexico City's Water Supply. Improving the Outlook for Sustainability*, Washington DC: National Academy Press.

National Statistics Office (1997), *1995 Census of Population*, Manila: National Statistics Office.

National Statistics Office (1998), *Philippine Statistical Yearbook*, Manila: National Statistics Office.

Nickson, Andrew (1996), *Urban Water Supply Sector Review*, Working Paper 7, International Development Department, School of Public Policy, Birmingham, UK: University of Birmingham.

Nickson, Andrew (1997), 'The Public-Private Mix in Urban Water Supply' in *International Review of Administrative Sciences,* 63, 165-186.

Noll, Roger, Mary M. Shirley and Simon Cowan (forthcoming), 'Reforming Urban Water Systems in Developing Countries' in Anne O. Krueger (ed.) *Economic Policy Reform*, Chicago: University of Chicago Press.

Panos (1998), *Liquid Assets. Is Water Privatisation the Answer to Access?* Media Briefing No. 29, London: The Panos Institute.

Peabody, N. (1991), *Water Policy Innovations in California: Water Resources Management in a Closing Water System*, Water Resources and Irrigation Policy Program Discussion Paper No.2, Arkansas: Winrock International Institute for Agricultural Development.

Pescuma, A. and M. E. Guaresti (1991), 'Gran Buenos Aires: Contaminación y Saneamiento' in *Medio Ambiente y Urbanización No. 37*, Buenos Aires: IIED América Latina.

Phillips, Charles F. (1993), *The Regulation of Public Utilities: Theory and Practice*, Arlington, Virginia: Public Utilities Reports.

Pickford, J. (1995), *Low-Cost Sanitation. A survey of practical experience*, Rugby, UK: Intermediate Technology Publications.

Raffiee, Kambiz (1993), 'Cost Analysis of Water Utilities: A Goodness-of-Fit Approach', *Atlantic Economic Journal,* 21 (3), 18-29.

Rees, Judith A. (1998), *Regulation and Private Participation in the Water and Sanitation Sector*, Global Water Partnership, Technical Advisory Committee Background Paper No. 1. Stockholm: Global Water Partnership.

Rivera, Daniel (1996), *Private Sector Participation in Water Supply and Sanitation: Lessons from Six Developing Countries*, World Bank Directions in Development, Washington DC: World Bank.

Roemer, Andrés (1997), *Derecho y Economía: Políticas Públicas del Agua*, Mexico: Centro de Investigación y Docencia Económicas, Sociedad Mexicana de Geografía y Estadística y Miguel Angel Porrúa.

Rondinelli, Dennis A. (1988), 'Increasing the Access of the Poor to Urban Services: Problems, Policy Alternatives and Organisational Choices' in D.-A. Rondinelli and G.-S. Cheema (eds) *Urban Services in Developing Countries: Public and Private Roles in Urban Development*, London: MacMillan.

Roth, Gabriel (1987), *The Private Provision of Public Services in Developing Countries*, Oxford: Oxford University Press.

Saade, Lilian (1993), *Perspectivas de la Participación de la Iniciativa Privada en el Servicio de Agua Potable en México*, tesis de licenciatura en Economía, Mexico: ITAM.

Saade, Lilian (1997), 'Toward more Efficient Urban Water Management in Mexico', *Water International,* **22** (3), 153-158.

Saade, Lilian (1998), 'New Strategy in Urban Water Management in Mexico: the Case of Mexico's Federal District', *Natural Resources Forum,* **22** (3), 185-192.

Sakho, M. (1998), *Current Situation in the Water Sector (water and development in the Ivory Coast)*, unpublished, Côte d'Ivoire: Direction de l'Eau.

Sappington, David E. M. (1996), 'Principles of Regulatory Policy Design' in A. Mody (ed.) *Infrastructure Delivery: Private Initiative and the Public Good.* Washington DC: World Bank Economic Development Institute.

SEMARNAP (1996), *Programa Hidráulico 1995-2000*, Mexico: Secretaría de Medio Abiente, Recursos Naturales y Pesca.

Sen, Amartaya (1983), 'Poor, Relatively Speaking', *Oxford Economic Papers,* **35**: 153-169.

Serageldin, Ismail (1994), *Water Supply, Sanitation and Environmental Sustainability*, Directions in Development, Washington DC: World Bank.

Shirley, Mary (1998), *Reforming Urban Water Systems: a Tale of Four Cities*, paper presented at the conference 'Regulation in Post-Privatization Environments: The Latin American Experience', 21-22 May 1998, Buenos Aires.

Silva, Gisele, Nicola Tynan and Yesim Yilmaz (1998), 'Private Participation in the Water and Sewerage Sector - Recent Trends' in *Public Policy for the Private Sector*, World Bank Note No. 147, Washington DC: World Bank.

SODECI (1996), *Rapport d'activité - SODECI*. Côte d'Ivoire: SODECI.

SODECI (1997), *Tarification de l'eau - SODECI*, Côte d'Ivoire: SODECI.

SODECI (1998), *Africa Initiative 2000*, Progress Report September 1998, Côte d'Ivoire: SODECI.

Solo, Maria Tovo (1998), *Small Scale Entrepreneurs in the Urban Water and Sanitation Markets*, UNDP/World Bank Water and Sanitation Programme draft mimeo, Washington DC: World Bank.

SOS (1998), *Memoria de Gestión 1994 - 1997*. Mexico: Comisión de Aguas del Distrito Federal, Secretaría de Obras y Servicios.

Teeples, R. and D. Glyer (1987), 'Cost of Water Delivery Systems: Specification and Ownership Effects', *Review of Economics and Statistics*, Jan 1987, 399-408.

UNICEF/WHO (1996), *Water Supply and Sanitation Monitoring Report 1996*, Geneva: WHO/UNICEF Joint Monitoring Programme

Urban Age (1999), 'Delivering Water to Mexico City', *The Global City Magazine*, 6(3).

Whittington, D., A. Okoarafor, A. Okore and A. McPhail (1991), 'A Study of Water Vending and Willingness to Pay for Water in Onitsha, Nigeria', *World Development*, 19 (2/3), 179-198.

Whittington, D., D. T. Lauria, A. Wright, K. Choe, J. Hughes and V. Swarna (1992), 'Household Demand for Improved Sanitation Services in Kumasi, Ghana: A Contingent Valuation Study', *Water Resources Research*, **29** (6), 1539-1560.

WHO (1996), *Creating Healthy Cities in the 21st Century*, Background Paper prepared for the Dialogue on Health in Human Settlements for Habitat II, Geneva: World Health Organization.

World Bank (1993a), *Philippines: Water Supply Sector Reform Study*, Final Report by Tasman Economic Research PTY Ltd., Washington DC: World Bank.

World Bank (1993b), *Water Resources Management*, Washington DC: World Bank.

World Bank (1994), *World Development Report: Infrastructure for Development*, Washington DC: World Bank.

World Bank (1995), *La Contaminación Ambiental en la Argentina: Problemas y Opciones*, Washington DC: World Bank.

World Bank (1996), *World Development Report*, Washington DC: World Bank.

World Bank (1998), *World Bank Viewpoint, Note No. 147*, Washington DC: World Bank.

World Resources Institute (1996-97), *World Resources 1996/97*, Oxford: Oxford University Press.

Wright, A. M. (1997), *Toward a Strategic Sanitation Approach: Improving the Sustainability of Urban Sanitation in Developing Countries*, Washington DC: World Bank.

Young, Robert A., Bruno Z. Haman, Danilo M. Cablyan, and Rolando M. Maloles (1996), *Water Management Allocation and Options: Angat River System*, Final Report of TA No. 2417-PHI, Manila: Asian Development Bank.

Zadi, K. M. 1995. *Participation du Secteur Privé dans la Fourniture des Services d'Alimentation en Eau et d'Assainissement: le Cas de la*

SODECI. Communication au séminaire Banque Mondiale Indonésie, Washington DC: World Bank.

Index